SOLUTIONS MANUAL FOR
RESONANT POWER CONVERTERS

SOLUTIONS MANUAL FOR
RESONANT POWER CONVERTERS

MARIAN K. KAZIMIERCZUK
Wright State University

DARIUSZ CZARKOWSKI
University of Florida

A Wiley-Interscience Publication
JOHN WILEY & SONS, INC.
New York • Chichester • Brisbane • Toronto • Singapore

Library of Congress Cataloging in Publication Data:

ISBN 0-471-12849-X (paper)

CONTENTS

Chapter 2

2.1 Compare the efficiency of the Class D current–driven center–tapped rectifier of Fig. 2.6 with *pn* junction diodes and Schottky diodes for $V_O = 5$ V and $I_O = 10$ A. The *pn* junction diode has $V_F = 0.8$ V and $R_F = 75$ mΩ. The Schottky diode has $V_F = 0.4$ V and $R_F = 25$ mΩ. Assume that in both cases the ESR of the filter capacitor is $r_C = 20$ mΩ and the transformer efficiency is $\eta_{tr} = 96\%$.

The load resistance is

$$R_L = \frac{V_O}{I_O} = \frac{5}{10} = 0.5 \ \Omega.$$

The efficiency is given by (2.71)

$$\eta_R = \frac{\eta_{tr}}{1 + \frac{V_F}{V_O} + \frac{\pi^2 R_F}{8R_L} + \frac{r_C}{R_L}\left(\frac{\pi^2}{8} - 1\right)}.$$

For the rectifier with *pn* junction diodes

$$\eta_R = \frac{0.96}{1 + \frac{0.8}{5} + \frac{\pi^2 \times 0.075}{8 \times 0.5} + \frac{0.02}{0.5}\left(\frac{\pi^2}{8} - 1\right)} = 70.88\%.$$

For the rectifier with Schottky diodes

$$\eta_R = \frac{0.96}{1 + \frac{0.4}{5} + \frac{\pi^2 \times 0.025}{8 \times 0.5} + \frac{0.02}{0.5}\left(\frac{\pi^2}{8} - 1\right)} = 83.4\%.$$

2.2 Derive an equation for the power factor of the transformerless version of the Class D current–driven half–wave rectifier. Compare the result with the expression for the transformer version of the rectifier.

The input voltage v_R of the transformerless Class D current–driven half–wave rectifier is approximately a square wave with a high level V_O and a zero low level. The rms value of v_R is

$$V_{Rrms} = \sqrt{\frac{1}{2\pi}\int_0^{2\pi} v_R^2 d(\omega t)} = \sqrt{\frac{V_O^2}{2\pi}\int_0^{\pi} d(\omega t)} = \frac{V_O}{\sqrt{2}}$$

the amplitude of the fundamental component of the input voltage is

$$V_{R1m} = \frac{1}{\pi} \int_0^{2\pi} v_R sin\omega t d(\omega t) = \frac{1}{\pi} \int_0^{\pi} V_O sin\omega t d(\omega t) = \frac{2V_O}{\pi}$$

the rms value of the fundamental component of the input voltage is

$$V_{R1rms} = \frac{V_{R1m}}{\sqrt{2}} = \frac{\sqrt{2}V_O}{\pi}$$

and the power factor is

$$PF = \frac{I_{Rrms}V_{R1rms}}{I_{Rrms}V_{Rrms}} = \frac{V_{R1rms}}{V_{Rrms}} = \frac{2}{\pi} \approx 0.64.$$

For the transformer version of the Class D current–driven half–wave rectifier, the power factor is given by (2.56) and is $\sqrt{2}$ times higher than that derived above.

2.3 Calculate the efficiency η_R, the voltage transfer function M_{VR}, and the input resistance R_i for a Class D bridge rectifier of Fig. 2.9 at $V_O = 5$ V and $I_O = 20$ A. The rectifier employs Schottky diodes with $V_F = 0.4$ V and $R_F = 0.025$ Ω and a filter capacitor C_f with $r_C = 20$ mΩ. The transformer turns ratio is $n = 5$. Assume the transformer efficiency $\eta_{tr} = 96\%$.

The load resistance is

$$R_L = \frac{V_O}{I_O} = \frac{5}{20} = 0.25 \ \Omega.$$

The efficiency is given by (2.96)

$$\eta_R = \frac{\eta_{tr}}{1 + \frac{2V_F}{V_O} + \frac{\pi^2 R_F}{4R_L} + \frac{r_C}{R_L}(\frac{\pi^2}{8} - 1)]}$$

$$= \frac{0.96}{1 + \frac{2\times0.4}{5} + \frac{\pi^2\times0.025}{4\times0.25} + \frac{0.02}{0.25}(\frac{\pi^2}{8} - 1)]} = 67.35\%.$$

From (2.97), the input resistance is

$$R_i = \frac{8n^2 R_L}{\pi^2 \eta_R} = \frac{8 \times 5^2 \times 0.25}{\pi^2 \times 0.6735} = 7.52 \ \Omega.$$

The voltage transfer function is calculated using (2.98)

$$M_{VR} = \frac{\pi \eta_R}{2\sqrt{2}n} = \frac{\pi \times 0.6735}{2\sqrt{2} \times 5} = 0.1496.$$

Note that the efficiency of the bridge rectifier is poor at low output voltage applications.

2.4 Prove that $v_c(0)$ in (2.39) is equal to $-\pi I_O/(2\omega C_f)$.

The voltage v_c is the ac component of the voltage across the filter capacitance C_f. This means that its average value is zero which can be described mathematically as

$$\frac{1}{2\pi}\int_0^{2\pi} v_c(\omega t)d(\omega t) = 0.$$

Substituting (2.39) to the above condition, one obtains

$$\int_0^{\pi}\left[\frac{\pi I_O}{\omega C_f}(1-\frac{\omega t}{\pi}-cos\omega t)+v_c(0)\right]d(\omega t)$$

$$+\int_{\pi}^{2\pi}\left[-\frac{I_O t}{C_f}+\frac{2\pi I_O}{\omega C_f}+v_c(0)\right]d(\omega t) = 0$$

$$\frac{\pi^2 I_O}{2\omega C_f}+\pi v_c(0)-\frac{3\pi^2 I_O}{2\omega C_f}+\frac{2\pi^2 I_O}{\omega C_f}+\pi v_c(0) = 0$$

and, eventually,

$$v_c(0) = -\frac{\pi I_O}{2\omega C_f}.$$

2.5 A half–wave rectifier, in which the maximum output current I_{Omax} is 5 A, operates at a switching frequency $f = 200$ kHz. The filter capacitance is $C_f = 100$ μF. What value of the ESR of the filter capacitor cannot be exceeded if it is specified that the ripple voltage V_r must be less than 0.3 V?

The maximum peak-to-peak ripple voltage across the filter capacitance is given by (2.41)

$$V_c = \frac{0.55 I_O}{fC_f} = \frac{0.55\times 5}{200\times 10^3\times 100\times 10^{-6}} = 137.5 \text{ mV}.$$

Hence, (2.43) leads to the maximum ripple voltage across the ESR of the filter capacitor as

$$V_{rESR} = V_r - V_c = 0.3-0.1375 = 162.5 \text{ mV}.$$

Using (2.42), the ESR of the filter capacitor cannot exceed the value

$$r_C = \frac{V_{rESR}}{\pi I_{Omax}} = \frac{0.1625}{\pi\times 5} = 10.4 \text{ m}\Omega.$$

2.6 Repeat Problem 2.5 for a transformer center–tapped rectifier.

From (2.84), the maximum peak-to-peak ripple voltage across the filter capacitance is

$$V_c = \frac{0.105 I_O}{f C_f} = \frac{0.105 \times 5}{200 \times 10^3 \times 100 \times 10^{-6}} = 26.25 \text{ mV}.$$

Since the obtained value of V_c is more than ten times smaller than maximum allowable V_r, it follows from (2.85) that V_{rESR} can be assumed to be equal to V_r. Hence, using (2.81), the ESR of the filter capacitor cannot be greater than

$$r_C = \frac{2 V_{rESR}}{\pi I_{Omax}} = \frac{2 \times 0.3}{\pi \times 5} = 38.2 \text{ m}\Omega.$$

As can be seen from this and previous problems, the condition for the ESR of the filter capacitor is almost four times less severe in the case of a transformer center–tapped rectifier than in the case of a half–wave rectifier.

2.7 The output filter of a Class D current–driven rectifier has the following parameters: $R_L = 2\ \Omega$, $r_C = 0.05\ \Omega$, $C_f = 47\ \mu F$, and $L_{ESL} = 20$ nH. Calculate the frequency at which the effect of ESL becomes significant. Is this frequency dependent upon R_L in a properly designed rectifier?

Referring to (2.115) and Fig. 2.15, the zeros of the input impedance of the output filter are

$$f_{z1,2} = \frac{1}{2\pi} \frac{r_C C_f \pm \sqrt{r_C^2 C_f^2 - 4 L_{ESL} C_f}}{2 L_{ESL} C_f}.$$

Increase of $|\,Z\,|$ due to the presence of L_{ESL} becomes significant at frequency

$$f_{z2} = \frac{1}{2\pi}$$

$$\times \frac{0.05 \times 47 \times 10^{-6} + \sqrt{0.05^2 \times 47^2 \times 10^{-12} - 4 \times 20 \times 10^{-9} \times 47 \times 10^{-6}}}{2 \times 20 \times 10^{-9} \times 47 \times 10^{-6}}$$

$$= 311.3 \text{ kHz}.$$

In a properly designed rectifier, i.e. with $r_C \ll R_L$, this frequency is independent of R_L because $f_{z2} \ll f_{p2}$.

2.8 A current–driven transformer center–tapped rectifier with output voltage $V_O = 3.3$ V and maximum output current $I_{Omax} = 10$ A employs Schottky diodes with $V_F = 0.3$ V and $R_F = 20$ mΩ, a filter capacitor with $C = 100$ μF and $r_C = 5$ mΩ, and a transformer with $n = 5$ and $\eta_{tr} = 96\%$. The operating frequency is $f = 150$ kHz. Find η_R, M_{VR}, R_i, and V_r.

The load resistance of the rectifier is

$$R_L = \frac{V_O}{I_O} = \frac{3.3}{10} = 0.33 \ \Omega$$

and the output power is

$$P_O = I_O V_O = 10 \times 3.3 = 33 \text{ W}.$$

The diode conduction loss can be calculated by substitution of (2.62) and (2.64) into (2.65)

$$P_D = P_{VF} + P_{RF} = \frac{V_F I_O}{2} + \frac{\pi^2 I_O^2 R_F}{16}$$

$$= \frac{0.3 \times 10}{2} + \frac{\pi^2 \times 10^2 \times 0.02}{16} = 2.73 \text{ W}.$$

From (2.68), one obtains the power loss in the filter capacitor

$$P_{rC} = r_C I_O^2 \left(\frac{\pi^2}{8} - 1 \right) = 0.005 \times 10^2 \left(\frac{\pi^2}{8} - 1 \right) = 0.12 \text{ W}.$$

Using (2.69),

$$P_C = 2P_D + P_{rC} = 5.46 + 0.12 = 5.58 \text{ W}.$$

From (2.71),

$$\eta_R = \frac{P_O \eta_{tr}}{P_O + P_C} = \frac{33 \times 0.96}{33 + 5.58} = 82.16\%.$$

The input resistance of the rectifier can be obtained from (2.72)

$$R_i = \frac{8n^2 R_L}{\pi^2 \eta_R} = \frac{8 \times 5^2 \times 0.33}{\pi^2 \times 0.8216} = 8.14 \ \Omega.$$

The voltage transfer function is calculated using (2.73)

$$M_{VR} = \frac{\pi \eta_R}{2\sqrt{2}n} = \frac{\pi \times 0.8216}{2\sqrt{2} \times 5} = 0.183.$$

From (2.81),

$$V_{rESR} = \frac{\pi r_C I_{Omax}}{2} = \frac{\pi \times 0.005 \times 10}{2} = 0.0785 \text{ V}$$

and from (2.84),

$$V_c = \frac{0.105 I_O}{f C_f} = \frac{0.105 \times 10}{150 \times 10^3 \times 100 \times 10^{-6}} = 0.07 \text{ V}.$$

Hence, using (2.85), the output ripple voltage can be estimated as

$$V_r = V_c + V_{rESR} = 0.07 + 0.0785 = 0.1485 \text{ V}.$$

2.9 Replace the diodes in the circuit given in Problem 2.8 with power SMP60N03–10L MOSFETs (Siliconix) to obtain an unregulated synchronous rectifier. The on–resistance of the power MOSFETs is r_{DS} = 10 mΩ, and the gate charge is Q_g = 100 nC. Calculate the rectifier efficiency and the drive power. Compare the efficiencies of the diode rectifier and the synchronous rectifier. What is the operating frequency at which the drive power is equal to the conduction loss?

Using (2.117), the gate drive power of each transistor can be obtained as

$$P_G = f Q_g V_{GSpp} = 150 \times 10^3 \times 100 \times 10^{-9} \times 10 = 0.15 \text{ W}.$$

From (2.119), the MOSFET conduction loss is

$$P_{rDS} = \frac{\pi^2 I_O^2 r_{DS}}{16} = \frac{\pi^2 \times 10^2 \times 0.01}{16} = 0.617 \text{ W}.$$

In Problem 2.8, the conduction loss in the filter capacitor was calculated to be P_{rC} = 0.12 W. Hence, (2.120) gives the conduction loss in the rectifier

$$P_C = 2 P_{rDS} + P_{rC} = 2 \times 0.617 + 0.12 = 1.354 \text{ W}.$$

From (2.122), the efficiency of the synchronous rectifier is

$$\eta_R = \frac{P_O \eta_{tr}}{P_O + P_C + 2 P_G} = \frac{33 \times 0.96}{33 + 1.354 + 0.3} = 91.42\%$$

and is almost 10% higher than the efficiency of the conventional rectifier of Problem 2.8.

The frequency at which the conduction losses and the gate-drive power are equal to each other can be found as

$$f = \frac{P_C}{2Q_g V_{GSpp}} = \frac{1.354}{2 \times 100 \times 10^{-9} \times 10} = 677 \text{ kHz.}$$

Chapter 3

3.1 Calculate the efficiency η_R, the voltage transfer function M_{VR}, and the input resistance R_i for a Class D half–wave rectifier of Fig. 3.1(a) at $V_O = 100$ V and $I_O = 1$ A. The operating frequency of the rectifier is $f = 100$ kHz. The rectifier employs p–n junction diodes with $V_F = 0.9$ V and $R_F = 0.04$ Ω. The value of the filter inductance is $L_f = 1$ mH. The dc ESR of the inductor is $r_{LF} = 0.1$ Ω, and the ac ESR of the inductor is $r_{Lf} = 1.85$ Ω. The ESR of the filter capacitor is $r_C = 50$ mΩ. The transformer turns ratio is $n = 2$. Assume the transformer efficiency $\eta_{tr} = 97\%$.

The load resistance is

$$R_L = \frac{V_O}{I_O} = \frac{100}{1} = 100 \ \Omega.$$

From (3.30), the efficiency is

$$\eta_R = \frac{\eta_{tr}}{1 + \frac{V_F}{V_O} + \frac{R_F + r_{LF}}{R_L} + a_{hw}^2 \frac{(r_{Lf} + r_C)R_L}{f^2 L_f^2}}$$

$$= \frac{0.97}{1 + \frac{0.9}{100} + \frac{0.04 + 0.1}{100} + 0.1808^2 \frac{(1.85 + 0.05) \times 100}{100^2 \times 10^6 \times 10^{-6}}} = 95.9\%.$$

The input resistance is given by (3.31)

$$R_i = \frac{\pi^2 n^2 R_L}{2\eta_R} = \frac{\pi^2 \times 2^2 \times 100}{2 \times 0.959} = 2058.3 \ \Omega$$

and the voltage transfer function is given by (3.34)

$$M_{VR} = \sqrt{\frac{\eta_R R_L}{R_i}} = \sqrt{\frac{0.959 \times 100}{2058.3}} = 0.216.$$

3.2 Repeat Problem 3.1 for the transformer center–tapped rectifier of Fig. 3.3(a).

The load resistance is

$$R_L = \frac{V_O}{I_O} = \frac{100}{1} = 100 \ \Omega.$$

The efficiency is given by (3.65)

$$\eta_R = \frac{\eta_{tr}}{1 + \frac{V_F}{V_O} + \frac{R_F + r_{LF}}{R_L} + a_{ct}^2 \frac{(r_{Lf} + r_C)R_L}{f^2 L_f^2}}$$

$$= \frac{0.97}{1 + \frac{0.9}{100} + \frac{0.04 + 0.1}{100} + 0.0377^2 \frac{(1.85 + 0.05) \times 100}{100^2 \times 10^6 \times 10^{-6}}} = 96\%.$$

From (3.66), the input resistance is

$$R_i = \frac{\pi^2 n^2 R_L}{8\eta_R} = \frac{\pi^2 \times 2^2 \times 100}{8 \times 0.96} = 514 \ \Omega.$$

Using (3.67), the voltage transfer function can be calculated as

$$M_{VR} = \sqrt{\frac{\eta_R R_L}{R_i}} = \sqrt{\frac{0.96 \times 100}{514}} = 0.4322.$$

3.3 Derive an expression for the power factor of the Class D transformerless voltage–driven half–wave rectifier. Explain why it is different than that for the transformer version of the half–wave rectifier.

The input current of the transformerless voltage–driven half–wave rectifier can be approximated by a square wave

$$i_R = \begin{cases} I_O, & \text{for} \quad 0 < \omega t \leq \pi \\ 0, & \text{for} \quad \pi < \omega t \leq 2\pi. \end{cases}$$

The amplitude of the fundamental component of i_R is

$$I_{R1m} = \frac{1}{\pi} \int_0^{2\pi} i_R sin\omega t d(\omega t) = \frac{I_O}{\pi} \int_0^{\pi} sin\omega t d(\omega t) = \frac{2I_O}{\pi}$$

and the rms value of the fundamental component is

$$I_{R1rms} = \frac{I_{R1m}}{\sqrt{2}} = \frac{\sqrt{2} I_O}{\pi}.$$

The rms value of the input current i_R is

$$I_{Rrms} = \sqrt{\frac{1}{2\pi} \int_0^{2\pi} i_R^2 d(\omega t)} = \sqrt{\frac{I_O^2}{2\pi} \int_0^{\pi} d(\omega t)} = \frac{I_O}{\sqrt{2}}.$$

Thus, from (3.11), the power factor is

$$PF = \frac{I_{R1rms}}{I_{Rrms}} = \frac{2}{\pi} \approx 0.64.$$

For the transformer half–wave rectifier, the rms value of the input current is given by (3.9) and is $\sqrt{2}$ times lower (neglecting the scaling factor n) than that of the transformerless rectifier. Therefore, the power factor of the transformerless rectifier is $\sqrt{2}$ times lower in comparison with the transformer version.

3.4 Calculate the efficiency η_R, the voltage transfer function M_{VR}, and the input resistance R_i for a Class D bridge rectifier of Fig. 3.5(a) at $V_O = 5$ V and $I_O = 20$ A. The operating frequency of the rectifier is 100 kHz. The rectifier employs Schottky diodes with $V_F = 0.4$ V and $R_F = 0.025$ Ω. The value of the filter inductance is $L_f = 1$ mH. The dc ESR of the inductor is $r_{LF} = 0.1$ Ω, and the ac ESR of the inductor is $r_{Lf} = 1.85$ Ω. A filter capacitor with $r_C = 50$ mΩ is employed. The transformer turns ratio is $n = 5$. Assume the transformer efficiency $\eta_{tr} = 0.96$.

The load resistance is

$$R_L = \frac{V_O}{I_O} = \frac{5}{20} = 0.25 \ \Omega.$$

The efficiency is given by (3.80)

$$\eta_R = \frac{\eta_{tr}}{1 + \frac{2V_F}{V_O} + \frac{2R_F + r_{LF}}{R_L} + a_b^2 \frac{r_{ac}R_L}{f^2 L_f^2}}$$

$$= \frac{0.96}{1 + \frac{2 \times 0.4}{5} + \frac{2 \times 0.025 + 0.1}{0.25} + 0.0377^2 \frac{(1.85 + 0.05) \times 0.25}{100^2 \times 10^6 \times 10^{-6}}} = 54.5\%.$$

Using (3.81), the input resistance is

$$R_i = \frac{\pi^2 n^2 R_L}{8 \eta_R} = \frac{\pi^2 \times 5^2 \times 0.25}{8 \times 0.545} = 14.15 \ \Omega.$$

From (3.82), the voltage transfer function is

$$M_{VR} = \sqrt{\frac{\eta_R R_L}{R_i}} = \sqrt{\frac{0.545 \times 0.25}{14.15}} = 0.098.$$

3.5 Derive Equation (3.43).

From (3.22) and (3.23), the minimum value of i_C is

$$i_C \left(arcsin \frac{1}{\pi} \right) = -1.73 \frac{V_O}{\omega L_f}$$

and the maximum value of i_C is

$$i_C \left(\pi - arcsin\frac{1}{\pi} \right) = \frac{\pi V_O}{2\omega L_f} = 1.73\frac{V_O}{\omega L_f}.$$

Hence,

$$I_{Cpp} = i_C \left(\pi - arcsin\frac{1}{\pi} \right) - i_C \left(arcsin\frac{1}{\pi} \right) = 3.46\frac{V_O}{\omega L_f}$$

and

$$I_{C1} = \frac{I_{Cpp}}{2} = 1.73\frac{V_O}{\omega L_f} = \frac{0.28V_O}{fL_f}.$$

3.6 Derive Equation (3.72).

Using (3.59) and (3.61), the peak-to-peak value of the current i_C is

$$I_{Cpp} = i_C \left(\pi - arcsin\frac{2}{\pi} \right) - i_C \left(arcsin\frac{2}{\pi} \right)$$

$$= 2i_C \left(\pi - arcsin\frac{2}{\pi} \right) = 2 \times 0.33\frac{V_O}{\omega L_f}.$$

Thus,

$$I_{C1} = \frac{I_{Cpp}}{2} = 0.33\frac{V_O}{\omega L_f} = \frac{0.05V_O}{fL_f}.$$

3.7 Show that $a_{hw} = (\sqrt{1/3 - 2/\pi^2})/2 = 0.1808$.

Substitution of (3.23) into (3.25) and relationships $V_O = I_O R_L$ and $\omega = 2\pi f$ yield

$$a_{hw} = \frac{1}{2\pi}$$

$$\times \sqrt{\frac{1}{2\pi} \left\{ \int_0^\pi \left[\pi(1 - cos\omega t) - \omega t - \frac{\pi}{2} \right]^2 d(\omega t) + \int_\pi^{2\pi} \left(-\omega t + 2\pi - \frac{\pi}{2} \right)^2 d(\omega t) \right\}}$$

and, after some tedious integration and algebra,

$$a_{hw} = \frac{1}{2\pi}\sqrt{\frac{1}{2\pi} \left(\frac{7\pi^3}{12} - 4\pi + \frac{\pi^3}{12} \right)} = \frac{\sqrt{\frac{2\pi^2}{3} - 4}}{2\sqrt{2\pi}} = 0.1808.$$

12

3.8 Show that $a_{ct} = (\sqrt{5/24 - 2/\pi^2})/2 = 0.0377$.

To get the value of a_{ct}, one should substitute (3.61) into (3.62) and solve the latter with respect to a_{ct} using relationships $V_O = I_O R_L$ and $\omega = 2\pi f$. This gives

$$a_{ct} = \frac{1}{2\pi}\sqrt{\frac{1}{\pi}\int_0^\pi \left[\frac{\pi}{2}(1 - cos\omega t) - \omega t\right]^2 d(\omega t)}$$

and, eventually,

$$a_{ct} = \frac{\sqrt{\frac{5\pi^2}{24} - 2}}{2\pi} = 0.0377.$$

Chapter 4

4.1 Derive Equation (4.14).

Using (4.7), the derivative of (4.10) is

$$\frac{1}{V_O}\frac{dv_D}{d(\omega t)} = \frac{1}{\omega C R_L}\left(1 - \cos\omega t - \frac{\sin\omega t}{\tan\phi}\right)$$

$$= \frac{1}{\omega C R_L}\left(1 - \frac{\cos\omega t \sin\phi + \sin\omega t \cos\phi}{\sin\phi}\right)$$

$$= \frac{1}{\omega C R_L}\left[1 - \frac{\sin(\omega t + \phi)}{\sin\phi}\right].$$

Voltage v_D reaches its minimum when

$$1 - \frac{\sin(\omega t_m + \phi)}{\sin\phi} = 0$$

or

$$\sin(\omega t_m + \phi) = \sin\phi.$$

Equations of the type $\sin(x) = a$ have two solutions per period: $x = \arcsin(a)$ and $x = \pi - \arcsin(a)$. In our case, the second solution has a physical meaning. Thus,

$$\omega t_m + \phi = \pi - \arcsin(\sin\phi)$$

and

$$\omega t_m = \pi - 2\phi.$$

Substitution of ωt_m into (4.10) gives

$$\frac{v_{Dmin}}{V_O} = \frac{1}{\omega C R_L}\left[\pi - 2\phi - \sin(\pi - 2\phi) + \frac{\cos(\pi - 2\phi) - 1}{\tan\phi}\right]$$

$$= \frac{1}{\omega C R_L}\left(\pi - 2\phi - \sin 2\phi - \frac{\cos 2\phi + 1}{\tan\phi}\right)$$

$$= \frac{1}{\omega C R_L}\left(\pi - 2\phi - 2\sin\phi\cos\phi - \frac{2\cos^2\phi}{\tan\phi}\right)$$

$$= \frac{1}{\omega C R_L}\left(\pi - 2\phi - \frac{2}{\tan\phi}\right).$$

Since $V_{DM} = -v_{Dmin}$, (4.14) has been derived.

13

4.2 Design a Class E low dv/dt rectifier of Fig. 4.1(a) with the following specifications: $V_O = 10$ V, $I_O = 0$ to 2 A, and $f = 500$ kHz.

The minimum load resistance is

$$R_{Lmin} = \frac{V_O}{I_{Omax}} = \frac{10}{2} = 5 \ \Omega.$$

The maximum output power is

$$P_{Omax} = V_O I_{Omax} = 10 \times 2 = 20 \ \text{W}.$$

Assume that $D = 0.5$ at $R_L = R_{Lmin}$. The maximum value of the diode peak current can be calculated using (4.33)

$$I_{DMmax} = 2.862 I_{Omax} = 2.862 \times 2 = 5.72 \ \text{A}.$$

From (4.34), the maximum value of the diode reverse peak voltage is

$$V_{DMmax} = 3.562 V_O = 3.562 \times 10 = 35.62 \ \text{V}.$$

The value of the capacitance C is given by (4.32)

$$C = \frac{1}{\pi \omega R_{Lmin}} = \frac{1}{2\pi^2 f R_{Lmin}} = \frac{1}{2\pi^2 \times 500 \times 10^3 \times 5} = 20.3 \ \text{nF}.$$

The voltage rating of the capacitor should be at least 40 V. Using (4.16) and (4.17), one can calculate the maximum value of the amplitude of the input current

$$I_{m(max)} = \frac{\sqrt{2} I_{Omax}}{M_{IR}} = \frac{\sqrt{2} \times 2}{0.7595} = 3.72 \ \text{A}.$$

4.3 Design a transformerless version of the Class E rectifier of Fig. 4.13(a) to meet the following specifications: $V_O = 10$ V, $I_O = 0$ to 2 A, and $f = 500$ kHz.

The minimum dc load resistance is

$$R_{Lmin} = \frac{V_O}{I_{Omax}} = \frac{10}{2} = 5 \ \Omega.$$

The maximum dc output power is

$$P_{Omax} = V_O I_{Omax} = 10 \times 2 = 20 \text{ W}.$$

Let us assume that $D = 0.5$ at the full load resistance. Thus, using (4.93),

$$L = \frac{R_{Lmin}}{\omega_o Q} = \frac{5}{0.3884 \times 2 \times \pi \times 500 \times 10^3} = 4.1 \ \mu\text{H}$$

and

$$C = \frac{Q}{\omega_o R_{Lmin}} = \frac{0.3884}{2 \times \pi \times 500 \times 10^3 \times 5} = 24.73 \text{ nF}.$$

From (4.94), the amplitude of the input voltage can be calculated as

$$V_m = \frac{\sqrt{2} V_O}{M_{VR}} = \frac{\sqrt{2} \times 10}{0.3684} = 38.39 \text{ V}.$$

Using (4.98) and (4.99), the maximum diode current and voltage are

$$I_{DM} = 2.777 I_{Omax} = 2.777 \times 2 = 5.56 \text{ A}$$

and

$$V_{DM} = 3.601 V_O = 3.601 \times 10 = 36.01 \text{ V}.$$

Chapter 5

5.1 Perform step-by-step integration in equation (5.24).

Using (5.22),

$$V_{Rim} = \frac{1}{\pi} \int_0^{2\pi} v_L sin(\omega t + \phi) d(\omega t)$$

$$= \frac{1}{\pi} \int_0^{2\pi D} V_O sin(\omega t + \phi) d(\omega t)$$

$$+ \frac{1}{\pi} \int_{2\pi D}^{2\pi} V_O \frac{cos(\omega t + \phi)}{cos\phi} sin(\omega t + \phi) d(\omega t)$$

$$= \frac{V_O}{\pi} [-cos(\omega t + \phi)] \Big|_0^{2\pi D} + \frac{V_O}{\pi cos\phi} \left[\frac{-cos^2(\omega t + \phi)}{2} \right] \Big|_{2\pi D}^{2\pi}.$$

Taking into account that $cos^2 x = 1 - sin^2 x$, one arrives after some algebra at

$$V_{Rim} = \frac{V_O}{\pi} \left[cos\phi - cos(2\pi D + \phi) + \frac{sin^2\phi - sin^2(2\pi D + \phi)}{2 cos\phi} \right].$$

5.2 Design a Class E low di/dt rectifier with a parallel inductor shown in Fig. 5.1(a). The following specifications should be met: $V_O = 20$ V, $R_L = 10$ Ω to infinity, $f = 0.5$ MHz, and $D = 0.5$ at the full load.

The full-power load resistance is $R_L = 10$ Ω. Using Table 5.1,

$$L = \frac{R_L}{3.1416\omega} = \frac{10}{3.1416 \times 2 \times \pi \times 0.5 \times 10^6} = 1.01 \ \mu H.$$

The maximum value of the dc load current is

$$I_{Omax} = \frac{V_O}{R_{Lmin}} = \frac{20}{10} = 2 \text{ A}.$$

From Table 5.1, $I_{DM}/I_O = 3.562$ at $D = 0.5$. Hence, the maximum value of the diode peak current is

$$I_{DMmax} = 3.562 I_{Omax} = 3.562 \times 2 = 7.124 \text{ A}.$$

From Table 5.1, $V_{DM}/V_O = 2.862$ at $D = 0.5$. Thus, the maximum value of the diode peak reverse voltage is

$$V_{DMmax} = 2.862V_O = 2.862 \times 20 = 57.24 \text{ V}.$$

Using (5.31) and Table 5.1, the amplitude of the input current is

$$I_{m(max)} = \frac{\sqrt{2}I_{Omax}}{M_{IR}} = \frac{\sqrt{2} \times 2}{0.2909} = 9.72 \text{ A}.$$

5.3 Design a Class E low di/dt rectifier with a series inductance that satisfies the following specifications: $V_O = 15$ V, $R_L = 10$ Ω to infinity, and $f = 200$ kHz. Assume that the maximum value of the diode ON duty cycle D_{max} cannot exceed 0.5.

The maximum duty cycle $D_{max} = 0.5$ occurs at the minimum resistance $R_{Lmin} = 10$ Ω. The maximum value of the load current is

$$I_{Omax} = \frac{V_O}{R_{Lmin}} = \frac{15}{10} = 1.5 \text{ A}.$$

From (5.46),

$$L = \frac{R_{Lmin}}{2\pi^2 f} = \frac{10}{2\pi^2 \times 200 \times 10^3} = 2.53 \text{ } \mu\text{H}.$$

Using Table 5.2, the peak values of the diode and inductor current and the diode reverse voltage are

$$I_{DM} = 3.5621 I_O = 3.5621 \times 1.5 = 5.34 \text{ A}$$

and

$$V_{DM} = 2.862V_O = 2.862 \times 15 = 42.9 \text{ V},$$

respectively. From (5.53) and Table 5.2, the rms value of the input voltage is

$$V_{rms} = \frac{V_O}{M_{VR}} = \frac{15}{0.7395} = 20.3 \text{ V}.$$

Using Table 5.2, the input resistance and the input inductance are

$$R_i = 1.7337 R_{Lmin} = 1.7337 \times 10 = 17.34 \text{ } \Omega$$

and

$$L_i = 4.726L = 4.726 \times 2.53 \times 10^{-6} = 11.96 \text{ } \mu\text{H}.$$

Chapter 6

6.1 A series–resonant circuit consists of an inductor $L = 84$ μH and a capacitor $C = 300$ pF. The ESRs of these components at the resonant frequency are $r_L = 1.4$ Ω and $r_C = 50$ mΩ, respectively. The load resistance is $R_i = 200$ Ω. The resonant circuit is driven by a sinusoidal voltage source whose amplitude is $V_m = 100$ V. Find the resonant frequency f_o, characteristic impedance Z_o, loaded quality factor Q_L, unloaded quality factor Q_o, quality factor of the inductor Q_{Lo}, and quality factor of the capacitor Q_{Co}.

From (6.8), the resonant frequency is

$$f_o = \frac{1}{2\pi\sqrt{LC}} = \frac{1}{2\pi\sqrt{84 \times 10^{-6} \times 300 \times 10^{-12}}} = 1 \text{ MHz}.$$

The characteristic impedance can be obtained using (6.9) as

$$Z_o = \sqrt{\frac{L}{C}} = \sqrt{\frac{84 \times 10^{-6}}{300 \times 10^{-12}}} = 529.2 \text{ } \Omega.$$

The parasitic resistance of the circuit is

$$r = r_L + r_C = 1.4 + 0.05 = 1.45 \text{ } \Omega$$

and the overall resistance is

$$R = R_i + r = 200 + 1.45 = 201.45 \text{ } \Omega.$$

Hence, from (6.10), the loaded quality factor is

$$Q_L = \frac{Z_o}{R} = \frac{529.2}{201.45} = 2.627.$$

The unloaded quality factor is given by (6.11)

$$Q_o = \frac{Z_o}{r} = \frac{529.2}{1.45} = 365.$$

Using (6.22) and (6.23), the quality factors of the inductor and the capacitor are

$$Q_{Lo} = \frac{Z_o}{r_L} = \frac{529.2}{1.4} = 378$$

and

$$Q_{Co} = \frac{Z_o}{r_C} = \frac{529.2}{0.05} = 10584.$$

6.2 For the resonant circuit given in Problem 6.1, find the reactive power of the inductor Q and the total true power P_R.

The quantities Q and P_R are defined at the resonant frequency. The amplitude of the current through the resonant circuit at this frequency is

$$I_m = \frac{V_m}{R}.$$

Hence, using (6.14) and (6.16), the reactive power of the inductor is

$$Q = \omega_o W_s = \omega_o \times \frac{1}{2}LI_m^2 = \frac{Z_o V_m^2}{2R^2} = \frac{529.2 \times 100^2}{2 \times 201.45^2} = 65.2 \text{ VA}.$$

From (6.19), the total true power is

$$P_R = \frac{V_m^2}{2R} = \frac{100^2}{2 \times 201.45} = 24.8 \text{ W}.$$

6.3 For the resonant circuit given in Problem 6.1, find the voltage and current stresses for the resonant inductor and the resonant capacitor. Calculate also the peak reactive power of the resonant components.

It can be assumed (see Problem 6.7) that the worst stresses occur at the resonant frequency which is

$$f_o = \frac{1}{2\pi\sqrt{LC}} = \frac{1}{2\pi\sqrt{84 \times 10^{-6} \times 300 \times 10^{-12}}} = 1 \text{ MHz}.$$

At this frequency, the amplitude of the current through the resonant circuit is

$$I_m = \frac{V_m}{R} = \frac{100}{201.45} = 0.4964 \text{ A}.$$

From (6.49), the amplitudes of the voltages across the resonant components are equal to each other and are

$$V_{Cm} = V_{Lm} = I_m Z_o = I_m\sqrt{\frac{L}{C}} = 0.4964\sqrt{\frac{84 \times 10^{-6}}{300 \times 10^{-12}}} = 262.7 \text{ V}.$$

The reactive powers for the resonant components are also equal and, using (6.14), can be calculated as

$$Q = \omega_o W_s = \frac{1}{2}\omega_o L I_m^2 = \frac{1}{2} \times 2 \times \pi \times 10^6 \times 84 \times 10^{-6} \times 0.4964^2 = 65 \text{ VA}.$$

6.4 Find the efficiency for the resonant circuit given in Problem 6.1. Is the efficiency dependent on the operating frequency?

The efficiency of the resonant circuit is given by (6.69)

$$\eta_r = \frac{R_i}{R_i + r} = \frac{R_i}{R_i + r_L + r_C} = \frac{200}{200 + 1.4 + 0.05} = 99.28\%.$$

This efficiency is independent of the operating frequency as long as the ESRs of the resonant components can be assumed independent of the frequency (which is not true when a wide range of frequencies is considered).

6.5 Write general expressions for the instantaneous energy stored in the resonant inductor $w_L(t)$ and in the resonant capacitor $w_C(t)$, as well as the total instantaneous energy stored in the resonant circuit $w_t(t)$. Sketch these waveforms for $f = f_o$. Explain briefly how the energy is transferred between the resonant components.

From (6.15) and (6.36), the energy stored in the inductor is

$$w_L(t) = \frac{1}{2}L I_m^2 \sin^2(\omega t - \psi) = \frac{L V_m^2 \cos^2\psi}{2R^2}\sin^2(\omega t - \psi)$$

where ψ is given by (6.26)

$$\psi = \arctan\left[\frac{\sqrt{\frac{L}{C}}}{R}\left(\frac{\omega}{\omega_o} - \frac{\omega_o}{\omega}\right)\right].$$

Likewise, using (6.15) and (6.47), the energy stored in the capacitor is

$$w_C(t) = \frac{1}{2}C V_{Cm}^2 \cos^2(\omega t - \psi) = \frac{V_m^2 \cos^2\psi}{2R^2\omega^2 C}\cos^2(\omega t - \psi).$$

The instantaneous energy stored in the circuit is

$$w_t(t) = w_L(t) + w_C(t)$$

$$= \frac{V_m^2 cos^2\psi}{2R^2}\left[Lsin^2(\omega t - \psi) + \frac{1}{\omega^2 C}cos^2(\omega t - \psi)\right].$$

At the resonant frequency, $\psi = 0$. Relationship $\omega_o L = 1/(\omega_o C)$ yields the expression for the instantaneous energy stored in the series resonant circuit at the resonant frequency

$$w_t(t) = \frac{V_m^2}{2R^2}[Lsin^2(\omega_o t) + Lcos^2(\omega_o t)] = \frac{V_m^2 L}{2R^2}.$$

This energy is constant for a given resonant circuit. Fig. P6.1 shows normalized total instantaneous energy $w_t \times 2R^2/(V_m^2 L) = 1$ stored in the series resonant circuit, normalized energy $w_L(t)/w_t = sin^2\omega_o t$ stored in the inductor, and normalized energy $w_C(t)/w_t = cos^2\omega_o t$ stored in the capacitor at the resonant frequency. It can be seen that as $w_L(t)$ reaches its maximum value at w_t, $w_C(t)$ reaches its minimum value at zero and vice versa.

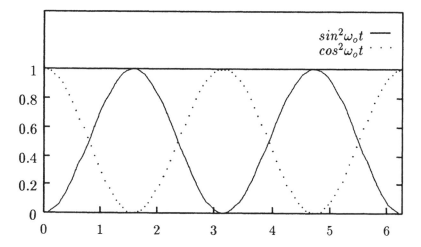

Figure P6.1: Normalized energies stored in a series resonant circuit at the resonant frequency.

6.6 A Class D half–bridge inverter is supplied by a dc voltage source of 350 to 400 V. Find the voltage stresses of the switches. Repeat the same problem for a Class D full–bridge inverter.

For the half-bridge inverter, the voltage stress of the switches is given by (6.45) and is

$$V_{SM} = V_{Imax} = 400 \text{ V}.$$

In the case of the full-bridge inverter, the voltage stress of the switches is given by (6.151) and is also

$$V_{SM} = V_{Imax} = 400 \text{ V.}$$

Voltage stresses of the switches for half-bridge and full-bridge topology are the same.

6.7 A series–resonant circuit, which consists of a resistance $R = 25 \ \Omega$, inductance $L = 100 \ \mu\text{H}$, and capacitance $C = 4.7 \ \text{nF}$, is driven by a sinusoidal voltage source $v = 100 sin\omega t$ (V). The operating frequency can be changed over a wide range. Calculate exactly the maximum voltage stresses for the resonant components. Compare the results with voltages across the inductance and the capacitance at the resonant frequency.

The amplitude of the input voltage v is $V_m = 100$ V, the resonant frequency is

$$f_o = \frac{1}{2\pi\sqrt{LC}} = \frac{1}{2\pi\sqrt{100 \times 10^{-6} \times 4.7 \times 10^{-9}}} = 232.15 \text{ kHz}$$

and the loaded quality factor is

$$Q_L = \frac{\sqrt{\frac{L}{C}}}{R} = \frac{\sqrt{\frac{100 \times 10^{-6}}{4.7 \times 10^{-9}}}}{25} = 5.835.$$

The amplitude of the voltage across the capacitor is given by (6.47)

$$V_{Cm} = \frac{V_m}{\frac{\omega}{\omega_o}\sqrt{\frac{1}{Q_L^2} + (\frac{\omega}{\omega_o} - \frac{\omega_o}{\omega})^2}}.$$

The derivative of V_{Cm} with respect to ω/ω_o is

$$\frac{dV_{Cm}}{d(\frac{\omega}{\omega_o})} = -\frac{V_m(\frac{\omega}{\omega_o} - \frac{\omega_o}{\omega})(1 + \frac{\omega_o^2}{\omega^2})}{\frac{\omega}{\omega_o}\left[\frac{1}{Q_L^2} + (\frac{\omega}{\omega_o} - \frac{\omega_o}{\omega})^2\right]^{\frac{3}{2}}} - \frac{V_m}{\frac{\omega^2}{\omega_o^2}\sqrt{\frac{1}{Q_L^2} + (\frac{\omega}{\omega_o} - \frac{\omega_o}{\omega})^2}}.$$

Setting this derivative to zero gives

$$\omega_{Cm} = \omega_o\sqrt{1 - \frac{1}{2Q_L^2}}$$

and

$$f_{Cm} = f_o\sqrt{1 - \frac{1}{2Q_L^2}} = 232.15 \times 10^3\sqrt{1 - \frac{1}{2 \times 5.835^2}} = 230.44 \text{ kHz.}$$

The maximum amplitude of the voltage across the capacitor is

$$V_{Cm}(\omega_{Cm}) = \frac{V_m Q_L}{\sqrt{1 - \frac{1}{4Q_L^2}}} = \frac{100 \times 5.835}{\sqrt{1 - \frac{1}{4 \times 5.835^2}}} = 585.65 \text{ V}.$$

Likewise, using (6.48), the amplitude of the voltage across the inductor is

$$V_{Lm} = \frac{V_m \frac{\omega}{\omega_o}}{\sqrt{\frac{1}{Q_L^2} + (\frac{\omega}{\omega_o} - \frac{\omega_o}{\omega})^2}}.$$

The derivative of V_{Lm} is

$$\frac{dV_{Lm}}{d(\frac{\omega}{\omega_o})} = -\frac{V_m \frac{\omega}{\omega_o}(\frac{\omega}{\omega_o} - \frac{\omega_o}{\omega})(1 + \frac{\omega_o^2}{\omega^2})}{[\frac{1}{Q_L^2} + (\frac{\omega}{\omega_o} - \frac{\omega_o}{\omega})^2]^{\frac{3}{2}}} - \frac{V_m}{\sqrt{\frac{1}{Q_L^2} + (\frac{\omega}{\omega_o} - \frac{\omega_o}{\omega})^2}}.$$

The maximum value of V_{Lm} occurs at

$$\omega_{Lm} = \frac{\omega_o}{\sqrt{1 - \frac{1}{2Q_L^2}}}$$

or

$$f_{Lm} = \frac{f_o}{\sqrt{1 - \frac{1}{2Q_L^2}}} = \frac{232.15 \times 10^3}{\sqrt{1 - \frac{1}{2 \times 5.835^2}}} = 233.87 \text{ kHz}.$$

The maximum amplitude of the voltage across the inductor is

$$V_{Lm}(\omega_{Lm}) = \frac{V_m Q_L}{\sqrt{1 - \frac{1}{4Q_L^2}}} = V_{Cm}(\omega_{Cm}) = 585.65 \text{ V}.$$

From (6.49), the amplitude of the voltage across the resonant components at the resonant frequency is

$$V_{Cm} = V_{Lm} = V_m Q_L = 100 \times 5.835 = 583.5 \text{ V}.$$

Thus, for sufficiently high Q_L, the assumption that the maximum voltage across the resonant components in a series resonant circuit occurs at the resonant frequency is valid for practical purposes. In this problem, for $Q_L = 5.835$, the error is

$$\frac{V_{Cm}(f_o) - V_{Cm}(f_{Cm})}{V_{Cm}(f_{Cm})} = \frac{583.5 - 585.65}{585.65} = -0.37\%.$$

6.8 Design a Class D half–bridge series resonant inverter that delivers to the load resistance power $P_O = 30$ W. The inverter is supplied from input voltage source $V_I = 180$ V. It is required that the switching frequency is $f = 210$ kHz. Neglect switching and drive–power losses.

Let us assume that $Q_L = 5$, $f/f_o = 1.05$, and $\eta_I = 96\%$ at the full load. From (6.43) and (6.69), the overall resistance of the inverter is

$$R = \frac{2V_I^2 \eta_I}{\pi^2 P_O[1 + Q_L^2(\frac{\omega}{\omega_o} - \frac{\omega_o}{\omega})^2]}$$

$$= \frac{2 \times 180^2 \times 0.96}{\pi^2 \times 30[1 + 5^2(1.05 - \frac{1}{1.05})^2]} = 169.7 \ \Omega.$$

The ac load resistance is

$$R_i = \eta_I R = 0.96 \times 169.7 = 162.9 \ \Omega.$$

The resonant frequency is

$$f_o = \frac{f}{\left(\frac{f}{f_o}\right)} = \frac{210 \times 10^3}{1.05} = 200 \text{ kHz}.$$

Using (6.10), the values of the resonant components are obtained as

$$L = \frac{Q_L R}{\omega_o} = \frac{5 \times 169.7}{2\pi \times 200 \times 10^3} = 675.2 \ \mu\text{H}$$

and

$$C = \frac{1}{\omega_o Q_L R} = \frac{1}{2\pi \times 200 \times 10^3 \times 5 \times 169.7} = 938 \text{ pF}.$$

The dc supply current is

$$I_I = \frac{P_I}{V_I} = \frac{P_O}{\eta_I V_I} = \frac{30}{0.96 \times 180} = 0.174 \text{ A}.$$

From from (6.45), the peak value of the switch voltage is

$$V_{SM} = V_I = 180 \text{ V}.$$

The peak value of the switch current is

$$I_m = \sqrt{\frac{2P_O}{R_i}} = \sqrt{\frac{2 \times 30}{162.9}} = 0.607 \text{ A}.$$

The maximum voltage stresses for the resonant components can be calculated using (6.49)

$$V_{Cm} = V_{Lm} = \frac{2V_I Q_L}{\pi} = \frac{2 \times 180 \times 5}{\pi} = 573 \text{ V}.$$

Chapter 7

7.1 The resonant circuit of Fig. P7.1 has the following parameters: $L = 200\ \mu\text{H}$, $C = 4.7$ nF, and $R_i = 500\ \Omega$. The circuit is driven by a variable frequency voltage source. Find the boundary frequency between the inductive and capacitive load for that source.

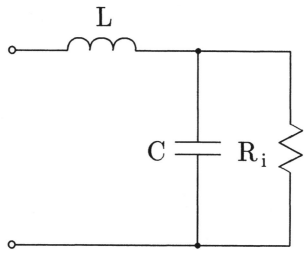

Figure P7.1: The resonant circuit of the inverter of Fig. 7.1(a).

The boundary between the inductive and capacitive load is at the resonant frequency which can be calculated from (7.15)

$$f_r = f_o\sqrt{1 - \frac{1}{Q_L^2}}.$$

Using (7.3), (7.4), and (7.5), f_r can be expressed as

$$f_r = \frac{1}{2\pi}\sqrt{\frac{1}{LC} - \frac{1}{R_i^2 C^2}}$$

$$= \frac{1}{2\pi}\sqrt{\frac{1}{200 \times 10^{-6} \times 4.7 \times 10^{-9}} - \frac{1}{500^2 \times 4.7^2 \times 10^{-18}}}$$

$$= 149.5\ \text{kHz}.$$

7.2 The resonant circuit of Fig. P7.1 has the following parameters: $L = 400\ \mu\text{H}$, $C = 10$ nF, and $R_i = 600\ \Omega$. The circuit is driven by a variable frequency voltage source $v = 100 sin \omega t$ (V). Calculate exactly the maximum voltage stresses for the resonant components. Compare the results with voltages across the inductor and the capacitor at the corner frequency.

From (7.5) and (7.4), the loaded quality factor of the circuit is

$$Q_L = \frac{R_i}{\sqrt{\frac{L}{C}}} = \frac{600}{\sqrt{\frac{400 \times 10^{-6}}{10 \times 10^{-9}}}} = 3.$$

The corner frequency is given by

$$f_o = \frac{1}{2\pi\sqrt{LC}} = \frac{1}{2\pi\sqrt{400 \times 10^{-6} \times 10 \times 10^{-9}}} = 79.58 \text{ kHz}.$$

The amplitude of the voltage across the capacitor is expressed by (7.51)

$$V_{Cm} = \frac{V_m}{\sqrt{[1 - (\frac{\omega}{\omega_o})^2]^2 + [\frac{1}{Q_L}(\frac{\omega}{\omega_o})]^2}}.$$

The derivative of V_{Cm} with respect to ω/ω_o is

$$\frac{dV_{Cm}}{d(\frac{\omega}{\omega_o})} = \frac{-V_m \left[\frac{2\frac{\omega}{\omega_o}}{Q_L^2} - 4\frac{\omega}{\omega_o}(1 - \frac{\omega^2}{\omega_o^2})^2\right]}{2\left[\frac{\omega^2}{\omega_o^2 Q_L^2} + (1 - \frac{\omega^2}{\omega_o^2})^2\right]^{\frac{3}{2}}}.$$

Setting this derivative to zero gives

$$\omega_{Cm} = \omega_o\sqrt{1 - \frac{1}{2Q_L^2}}$$

and

$$f_{Cm} = f_o\sqrt{1 - \frac{1}{2Q_L^2}} = 79.58 \times 10^3 \sqrt{1 - \frac{1}{2 \times 3^2}} = 77.34 \text{ kHz}.$$

The maximum amplitude of the voltage across the capacitor is

$$V_{Cm}(\omega_{Cm}) = \frac{V_m Q_L}{\sqrt{1 - \frac{1}{4Q_L^2}}} = \frac{100 \times 3}{\sqrt{1 - \frac{1}{4 \times 3^2}}} = 304.26 \text{ V}.$$

As was expected, the obtained above results are in full agreement with (7.38) and (7.39).

The amplitude of the voltage across the inductor is given by (7.53)

$$V_{Lm} = V_m \left(\frac{\omega}{\omega_o}\right) \sqrt{\frac{1 + (Q_L \frac{\omega}{\omega_o})^2}{Q_L^2[1 - (\frac{\omega}{\omega_o})^2]^2 + (\frac{\omega}{\omega_o})^2}}.$$

The derivative of V_{Lm} with respect to ω/ω_o is

$$\frac{dV_{Lm}}{d(\frac{\omega}{\omega_o})} = V_m \sqrt{\frac{1 + Q_L^2 \frac{\omega^2}{\omega_o^2}}{\frac{\omega^2}{\omega_o^2} + Q_L^2(1 - \frac{\omega^2}{\omega_o^2})^2}}$$

$$+ \frac{V_m \frac{\omega}{\omega_o} \left\{ \frac{2Q_L^2 \frac{\omega}{\omega_o}}{\frac{\omega^2}{\omega_o^2} + Q_L^2(1 - \frac{\omega^2}{\omega_o^2})^2} - \frac{(1 + Q_L^2 \frac{\omega^2}{\omega_o^2})[2\frac{\omega}{\omega_o} - 4Q_L^2 \frac{\omega}{\omega_o}(1 - \frac{\omega^2}{\omega_o^2})^2]}{[\frac{\omega^2}{\omega_o^2} + Q_L^2(1 - \frac{\omega^2}{\omega_o^2})^2]^2} \right\}}{2\sqrt{\frac{1 + Q_L^2 \frac{\omega^2}{\omega_o^2}}{\frac{\omega^2}{\omega_o^2} + Q_L^2(1 - \frac{\omega^2}{\omega_o^2})^2}}}.$$

After some tedious algebra, equation $dV_{Lm}/d(\omega/\omega_o) = 0$ becomes equivalent to

$$2Q_L^4 \frac{\omega^4}{\omega_o^4} - 2Q_L^4 \frac{\omega^2}{\omega_o^2} + Q_L^2 = 0$$

which results in

$$\omega_{Lm} = \omega_o \sqrt{\frac{1}{2} + \frac{\sqrt{2 + Q_L^2}}{2Q_L}}$$

and

$$f_{Lm} = f_o \sqrt{\frac{1}{2} + \frac{\sqrt{2 + Q_L^2}}{2Q_L}}$$

$$= 79.58 \times 10^3 \sqrt{\frac{1}{2} + \frac{\sqrt{2 + 3^2}}{2 \times 3}} = 81.65 \text{ kHz}.$$

The maximum amplitude of the voltage across the inductor is

$$V_{Lm}(\omega_{Lm}) = V_m \sqrt{\frac{1}{2} + \frac{\sqrt{2 + Q_L^2}}{2Q_L}}$$

$$\times \sqrt{\frac{Q_L(Q_L + 2Q_L^3 - \sqrt{2 + Q_L^2} + 2Q_L^2\sqrt{2 + Q_L^2})}{4Q_L^2 - 1}}$$

$$= 100 \sqrt{\frac{1}{2} + \frac{\sqrt{2 + 3^2}}{2 \times 3}} \sqrt{\frac{3(3 + 2 \times 3^3 - \sqrt{2 + 3^2} + 2 \times 3^2\sqrt{2 + 3^2})}{4 \times 3^2 - 1}}$$

$$= 319.9 \text{ V}.$$

From (7.52) and (7.54), the voltage stresses at the corner frequency are

$$V_{Cm} = V_m Q_L = 100 \times 3 = 300 \text{ V}$$

and

$$V_{Lm} = V_m \sqrt{Q_L^2 + 1} = 100\sqrt{3^2 + 1} = 316.23 \text{ V}.$$

It can be seen that the differences between exact and approximate calculations are very small. Those differences decrease with Q_L.

7.3 Derive equations (7.40) and (7.41).

From (7.35)

$$M_{Vr} = \frac{V_{Ri}}{V_{rms}} = \frac{1}{\sqrt{[1 - (\frac{f}{f_o})^2]^2 + [\frac{1}{Q_L}(\frac{f}{f_o})]^2}}.$$

Hence

$$\left[1 - \left(\frac{f}{f_o}\right)^2\right]^2 + \left[\frac{1}{Q_L}\left(\frac{f}{f_o}\right)\right]^2 = \frac{1}{M_{Vr}^2}.$$

Denoting $x = (f/f_o)^2$

$$x^2 - \left(\frac{1}{Q_L^2} - 2\right)x + 1 - \frac{1}{M_{Vr}^2} = 0.$$

Solving this quadratic equation with respect to x

$$x_{1,2} = \frac{2 - \frac{1}{Q_L^2} \pm \sqrt{(\frac{1}{Q_L^2} - 2)^2 - 4 + \frac{4}{M_{Vr}^2}}}{2}$$

$$x_{1,2} = 1 - \frac{1}{2Q_L^2} \pm \sqrt{\left(\frac{1}{2Q_L^2} - 1\right)^2 - 1 + \frac{1}{M_{Vr}^2}}.$$

Thus

$$\frac{f}{f_o} = \sqrt{1 - \frac{1}{2Q_L^2} + \sqrt{\left(1 - \frac{1}{2Q_L^2}\right)^2 + \frac{1}{M_{Vr}^2} - 1}}$$

or

$$\frac{f}{f_o} = \sqrt{1 - \frac{1}{2Q_L^2} - \sqrt{\left(1 - \frac{1}{2Q_L^2}\right)^2 + \frac{1}{M_{Vr}^2} - 1}}.$$

7.4 Show that $R_{icr} = 1/(\omega C)$ in (7.61).

The efficiency of the inverter is given by (7.60)

$$\eta_I = \frac{1}{1 + \frac{r}{Z_o Q_L}[1 + (\frac{\omega}{\omega_o})^2 Q_L^2]}.$$

The derivative of η_I with respect to Q_L is

$$\frac{d\eta_I}{dQ_L} = -\frac{\frac{2r(\frac{\omega}{\omega_o})^2}{Z_o} - \frac{r[1+(\frac{\omega}{\omega_o})^2 Q_L^2]}{Q_L^2 Z_o}}{\{1 + \frac{r}{Z_o Q_L}[1 + (\frac{\omega}{\omega_o})^2 Q_L^2]\}}.$$

Setting this derivative to zero results in

$$Q_{Lcr} = \frac{R_{icr}}{Z_o} = \frac{\omega_o}{\omega}.$$

Thus, using (7.3) and (7.4),

$$R_{icr} = \frac{Z_o \omega_o}{\omega} = \frac{\sqrt{\frac{L}{C}} \frac{1}{\sqrt{LC}}}{\omega} = \frac{1}{\omega C}.$$

7.5 Design a full-bridge Class D parallel resonant inverter to meet the following specifications: $V_I = 200$ V, $P_O = 75$ W, and $f = 100$ kHz. Assume the loaded quality factor at full load to be $Q_L = 2.5$ and that the inverter efficiency $\eta_I = 92\%$.

The dc supply power is

$$P_I = \frac{P_O}{\eta_I} = \frac{75}{0.92} = 81.52 \text{ W}$$

and the dc supply current is

$$I_I = \frac{P_I}{V_I} = \frac{81.52}{200} = 407.6 \text{ mA}.$$

Assuming that $f = f_r = 100$ kHz at full power and using (7.15), the corner frequency is

$$f_o = \frac{f_r}{\sqrt{1 - \frac{1}{Q_L^2}}} = \frac{100 \times 10^3}{\sqrt{1 - \frac{1}{2.5^2}}} = 109.1 \text{ kHz}.$$

From (7.98) and (7.46), one obtains the ac load resistance of the inverter as

$$R_i = \frac{V_{Ri}^2}{P_O} = \frac{8V_I^2\eta_I^2}{\pi^2 P_O\{[1-(\frac{\omega}{\omega_o})^2]^2 + [\frac{1}{Q_L}(\frac{\omega}{\omega_o})]^2\}}$$

$$= \frac{8 \times 200^2 \times 0.92^2}{\pi^2 \times 75[(1-\frac{100^2}{109.1^2})^2 + (\frac{100}{2.5\times109.1})^2]} = 2287.2 \ \Omega.$$

Using (7.5), the characteristic impedance is

$$Z_o = \frac{R_i}{Q_L} = \frac{1614.4}{2.5} = 914.9 \ \Omega.$$

From (7.4), the elements of the resonant circuit are

$$L = \frac{Z_o}{\omega_o} = \frac{914.9}{2\pi \times 109.1 \times 10^3} = 1.335 \ \text{mH}$$

and

$$C = \frac{1}{\omega_o Z_o} = \frac{1}{2\pi \times 109.1 \times 10^3 \times 914.9} = 1.59 \ \text{nF}.$$

From (7.100), the peak value of the switch current is

$$I_m = I_{SM} = \frac{4V_I\sqrt{Q_L^2+1}}{\pi Z_o} = \frac{4 \times 200\sqrt{2.5^2+1}}{\pi \times 914.9} = 0.75 \ \text{A}.$$

Using (7.105) and (7.107), the voltage stresses of the resonant components are obtained as

$$V_{Cm} = \frac{4V_I Q_L}{\pi} = \frac{4 \times 200 \times 2.5}{\pi} = 636.6 \ \text{V}$$

and

$$V_{Lm} = \frac{4V_I\sqrt{Q_L^2+1}}{\pi} = \frac{4 \times 200 \times \sqrt{2.5^2+1}}{\pi} = 685.6 \ \text{V}.$$

Chapter 8

8.1 Derive, step by step, the input impedance of the resonant circuit of Fig. 8.1. Compare your result to (8.9).

The impedances of the elements of the resonant circuit of Fig. 8.1 in the frequency domain are: $j\omega L$, $1/(j\omega C_1)$, $1/(j\omega C_2)$, and R_i. Let us denote the impedance of the series combination $L-C_1$ as Z_1 and the impedance of the parallel combination C_2-R_i as Z_2. Thus, using (8.2), (8.4), and (8.5)

$$Z_1 = j\omega L + \frac{1}{j\omega C_1} = jZ_o\frac{\omega}{\omega_o} + \frac{1}{j\omega C(1 + \frac{1}{A})}$$

$$= jZ_o\frac{\omega}{\omega_o} - jZ_o\frac{\omega_o}{\omega}\frac{A}{A+1} = jZ_o\left(\frac{\omega}{\omega_o} - \frac{\omega_o}{\omega}\frac{A}{A+1}\right)$$

and

$$Z_2 = R_i \parallel \frac{1}{j\omega C_2} = \frac{\frac{R_i}{j\omega C_2}}{R_i + \frac{1}{j\omega C_2}} = \frac{R_i}{1 + j\omega C_2 R_i}$$

$$= \frac{R_i}{1 + jR_i\omega_o C\frac{\omega}{\omega_o}(1 + A)} = \frac{R_i}{1 + jQ_L\frac{\omega}{\omega_o}(1 + A)}.$$

The input impedance of the resonant circuit is

$$\mathbf{Z} = Z_1 + Z_2 = jZ_o\left(\frac{\omega}{\omega_o} - \frac{\omega_o}{\omega}\frac{A}{A+1}\right) + \frac{R_i}{1 + jQ_L\frac{\omega}{\omega_o}(1 + A)}$$

$$= \frac{jZ_o(\frac{\omega}{\omega_o} - \frac{\omega_o}{\omega}\frac{A}{A+1})[1 + jQ_L\frac{\omega}{\omega_o}(1 + A)] + R_i}{1 + jQ_L\frac{\omega}{\omega_o}(1 + A)}$$

$$= \frac{jZ_o(\frac{\omega}{\omega_o} - \frac{\omega_o}{\omega}\frac{A}{A+1}) - R_i[(1 + A)\frac{\omega^2}{\omega_o^2} - A] + R_i}{1 + jQ_L\frac{\omega}{\omega_o}(1 + A)}$$

$$= \frac{R_i\{(1 + A)[1 - (\frac{\omega}{\omega_o})^2] + j\frac{1}{Q_L}(\frac{\omega}{\omega_o} - \frac{\omega_o}{\omega}\frac{A}{A+1})\}}{1 + jQ_L\frac{\omega}{\omega_o}(1 + A)}.$$

8.2 The resonant circuit of the inverter of Fig. 8.1 with $L = 500\ \mu H$, $C_1 = C_2 = 4.7$ nF, and $R_i = 600\ \Omega$ is driven by a sinusoidal voltage source $v = 100sin(840 \times 10^3 t)$. What is the amplitude of the voltage across the ac load R_i in this circuit?

From (8.1) and (8.2), the ratio of the capacitances is $A = C_2/C_1 = 1$ and the equivalent capacitance is

$$C = \frac{C_2}{1+A} = \frac{4.7 \times 10^{-9}}{1+1} = 2.35 \text{ nF}.$$

The corner frequency is given by (8.3)

$$\omega_o = \frac{1}{\sqrt{LC}} = \frac{1}{\sqrt{500 \times 10^{-6} \times 2.35 \times 10^{-9}}} = 922.5 \times 10^3 \frac{1}{\text{s}}.$$

Thus, the ratio ω/ω_o is $\omega/\omega_o = 840/922.5 = 0.91$. Using (8.5), the loaded quality factor Q_L is calculated as

$$Q_L = \frac{R_i}{\omega_o L} = \frac{600}{922.5 \times 10^3 \times 500 \times 10^{-6}} = 1.3.$$

The amplitude of the voltage across the ac load R_i is equal to the amplitude of the voltage across capacitor C_2 given by (8.34)

$$V_{Rim} = \sqrt{2}V_{Ri} = \frac{V_m}{\sqrt{(1+A)^2[1-(\frac{\omega}{\omega_o})^2]^2 + [\frac{1}{Q_L}(\frac{\omega}{\omega_o} - \frac{\omega_o}{\omega}\frac{A}{A+1})]^2}}$$

$$= \frac{100}{\sqrt{(1+1)^2[1-0.91^2]^2 + [\frac{1}{1.3}(0.91 - \frac{1}{0.91}\frac{1}{1+1})]^2}} = 226.4 \text{ V}.$$

8.3 Calculate the voltage stresses for the resonant components of the circuit from Problem 8.2.

The parameters A, ω/ω_o, and Q_L of the resonant circuit as well as the amplitude of the voltage across capacitor C_2 are calculated in the solution to the previous problem. The amplitude of the voltage across the resonant inductor can be calculated from (8.32)

$$V_{Lm} = \frac{V_m(\frac{\omega}{\omega_o})}{Q_L}\sqrt{\frac{1 + [Q_L(\frac{\omega}{\omega_o})(1+A)]^2}{(1+A)^2[1-(\frac{\omega}{\omega_o})^2]^2 + \frac{1}{Q_L^2}(\frac{\omega}{\omega_o} - \frac{\omega_o}{\omega}\frac{A}{A+1})^2}}$$

$$= \frac{100 \times 0.91}{1.3}\sqrt{\frac{1 + [1.3 \times 0.91(1+1)]^2}{(1+1)^2[1-0.91^2]^2 + [\frac{1}{1.3}(0.91 - \frac{1}{0.91}\frac{1}{1+1})]^2}}$$

$$= 407.1 \text{ V}.$$

Using (8.33), the amplitude of the voltage across the resonant capacitor C_1 is

$$V_{C1m} = \frac{V_m A}{Q_L(\frac{\omega}{\omega_o})(1+A)} \sqrt{\frac{1 + [Q_L(\frac{\omega}{\omega_o})(1+A)]^2}{(1+A)^2[1 - (\frac{\omega}{\omega_o})^2]^2 + \frac{1}{Q_L^2}(\frac{\omega}{\omega_o} - \frac{\omega_o}{\omega}\frac{A}{A+1})^2}}$$

$$= \frac{100 \times 1}{1.3 \times 0.91(1+1)} \sqrt{\frac{1 + [1.3 \times 0.91(1+1)]^2}{(1+1)^2[1 - 0.91^2]^2 + [\frac{1}{1.3}(0.91 - \frac{1}{0.91}\frac{1}{1+1})]^2}}$$

$$= 245.78 \text{ V}.$$

8.4 Derive equation (8.41).

The efficiency of the inverter is given by (8.40)

$$\eta_I = \frac{1}{1 + \frac{r}{Z_o Q_L}\{1 + [Q_L(\frac{\omega}{\omega_o})(1+A)]^2\}}.$$

The derivative of η_I with respect to Q_L is

$$\frac{d\eta_I}{dQ_L} = -\frac{\frac{2r[\frac{\omega}{\omega_o}(1+A)]^2}{Z_o} - \frac{r\{1 + Q_L^2[\frac{\omega}{\omega_o}(1+A)]^2\}}{Z_o Q_L^2}}{\{1 + \frac{r}{Z_o Q_L}[1 + Q_L^2(\frac{\omega}{\omega_o})^2(1+A)^2]\}^2}.$$

Setting this derivative to zero gives (8.41)

$$Q_L = \frac{1}{\left(\frac{\omega}{\omega_o}\right)(1+A)}.$$

8.5 Design a full-bridge Class D series-parallel inverter of Fig. 8.16 to meet the following specifications: $V_I = 250$ V, $f_o = 100$ kHz, $R_{imin} = 200$ Ω, and $P_{Rimax} = 85$ W. Assume the inverter efficiency $\eta_I = 90\%$, $A = 1$, and $f/f_o = 0.85$.

The maximum dc input power is

$$P_{Imax} = \frac{P_{Rimax}}{\eta_I} = \frac{85}{0.9} = 94.4 \text{ W}$$

and the maximum value of the dc input current is

$$I_{Imax} = \frac{P_{Imax}}{V_I} = \frac{94.4}{250} = 0.38 \text{ A}.$$

The voltage transfer function of the inverter is

$$M_{VIa} = \frac{V_{Ri}}{V_I} = \frac{\sqrt{P_{Ri}R_i}}{V_I} = \frac{\sqrt{85 \times 200}}{250} = 0.5215.$$

Using (8.28) and (8.67), the loaded quality factor can be calculated as

$$Q_L = \frac{\frac{\omega}{\omega_o} - \frac{\omega_o}{\omega}\frac{A}{A+1}}{\sqrt{\frac{8\eta_I^2}{\pi^2 M_{VIa}^2} - (1+A)^2[1 - (\frac{\omega}{\omega_o})^2]^2}}$$

$$= \frac{0.85 - \frac{1}{0.85 \times 2}}{\sqrt{\frac{8 \times 0.9^2}{\pi^2 \times 0.5215^2} - 2^2(1 - 0.85^2)^2}} = 0.18.$$

From (8.5) and (8.2), the values of the resonant components are

$$L = \frac{R_{imin}}{\omega_o Q_L} = \frac{200}{2 \times \pi \times 100 \times 10^3 \times 0.18} = 1.768 \text{ mH}$$

$$C = \frac{Q_L}{\omega_o R_{imin}} = \frac{0.18}{2 \times \pi \times 100 \times 10^3 \times 200} = 1.43 \text{ nF}$$

$$C_1 = C\left(1 + \frac{1}{A}\right) = 2C = 2.86 \text{ nF}$$

$$C_2 = C(1 + A) = 2C = 2.86 \text{ nF}.$$

Using (8.16), the resonant frequency is

$$f_r = f_o\sqrt{\frac{Q_L^2(1+A)^2 - 1 + \sqrt{[Q_L^2(1+A)^2 - 1]^2 + 4Q_L^2 A(1+A)}}{2Q_L^2(1+A)^2}}$$

$$= 100 \times 10^3 \sqrt{\frac{4 \times 0.18^2 - 1 + \sqrt{(4 \times 0.18^2 - 1)^2 + 8 \times 0.18^2}}{8 \times 0.18^2}}$$

$$= 72.96 \text{ kHz}.$$

Since $f > f_r$, the switches are loaded inductively.

Using (8.69), the amplitude of the current through the resonant circuit can be calculated as

$$I_m = \sqrt{\frac{2P_{Ri}\{1 + [Q_L(\frac{\omega}{\omega_o})(1+A)]^2\}}{R_i}}$$

$$= \sqrt{\frac{2 \times 85\{1 + [0.18 \times 0.85 \times (1+1)]^2\}}{200}} = 0.97 \text{ A}.$$

Hence, the maximum value of the current through the switches is $I_{SM} = I_m = 0.97$ A.

The peak values of the voltages across the reactive components can be obtained from (8.70), (8.71), and (8.72) as

$$V_{Lm} = \omega L I_m = 2\pi \times 85 \times 10^3 \times 1.768 \times 10^{-3} \times 0.97 = 915.9 \text{ V}$$

$$V_{C1m} = \frac{I_m}{\omega C_1} = \frac{0.97}{2\pi \times 85 \times 10^3 \times 2.86 \times 10^{-9}} = 635.1 \text{ V}$$

and

$$V_{C2m} = \sqrt{2} V_{Ri} = \sqrt{2 P_{Ri} R_i} = \sqrt{2 \times 85 \times 200} = 184.4 \text{ V}.$$

Chapter 9

9.1 Derive step by step the input impedance of the resonant circuit of Fig. 9.1. Compare your result to (9.9).

The impedances of the elements of the resonant circuit of Fig. 9.1 in the frequency domain are: $1/(j\omega C_2)$, $j\omega L_1$, $j\omega L_2$, and R_i. Let us denote the impedance of the series combination C–L_1 as Z_1 and the impedance of the parallel combination L_2–R_i as Z_2. Hence, using (9.2), (9.4), and (9.5)

$$Z_1 = j\omega L_1 + \frac{1}{j\omega C} = j\omega_o \frac{L}{1 + \frac{1}{A}\omega_o}\frac{\omega}{} - \frac{j\omega_o}{\omega\omega_o C}$$

$$= jZ_o \frac{\omega}{\omega_o}\frac{A}{A+1} - jZ_o\frac{\omega_o}{\omega} = jZ_o\left(\frac{\omega}{\omega_o}\frac{A}{A+1} - \frac{\omega_o}{\omega}\right)$$

and

$$Z_2 = R_i \parallel j\omega L_2 = \frac{j\omega R_i L_2}{R_i + j\omega L_2} = \frac{R_i}{1 + \frac{R_i}{j\omega L_2}}$$

$$= \frac{R_i}{1 - j\frac{R_i}{\omega_o L}\frac{\omega_o}{\omega}(1+A)} = \frac{R_i}{1 - jQ_L\frac{\omega_o}{\omega}(1+A)}.$$

The input impedance of the resonant circuit is

$$\mathbf{Z} = Z_1 + Z_2 = jZ_o\left(\frac{\omega}{\omega_o}\frac{A}{A+1} - \frac{\omega_o}{\omega}\right) + \frac{R_i}{1 - jQ_L\frac{\omega_o}{\omega}(1+A)}$$

$$\frac{jZ_o(\frac{\omega}{\omega_o}\frac{A}{A+1} - \frac{\omega_o}{\omega})[1 - jQ_L\frac{\omega_o}{\omega}(1+A)] + R_i}{1 - jQ_L\frac{\omega_o}{\omega}(1+A)}$$

$$= \frac{jZ_o(\frac{\omega}{\omega_o}\frac{A}{A+1} - \frac{\omega_o}{\omega}) + R_i[A - (1+A)\frac{\omega_o^2}{\omega^2}] + R_i}{1 - jQ_L\frac{\omega_o}{\omega}(1+A)}$$

$$= \frac{R_i\{(1+A)[1 - (\frac{\omega_o}{\omega})^2] + j\frac{1}{Q_L}(\frac{\omega}{\omega_o}\frac{A}{A+1} - \frac{\omega_o}{\omega})\}}{1 - jQ_L(\frac{\omega_o}{\omega})(1+A)}.$$

9.2 Show that the argument of the input impedance of the resonant circuit of Fig. 9.1 is given by (9.11).

Let us multiply the numerator and denominator of the input impedance by $[1 + jQ_L\frac{\omega_o}{\omega}(1+A)]$.

$$\mathbf{Z} = \frac{R_i\{(1+A)[1-(\frac{\omega_o}{\omega})^2]+j\frac{1}{Q_L}(\frac{\omega}{\omega_o}\frac{A}{A+1}-\frac{\omega_o}{\omega})\}[1+jQ_L(\frac{\omega_o}{\omega})(1+A)]}{[1-jQ_L(\frac{\omega_o}{\omega})(1+A)][1+jQ_L(\frac{\omega_o}{\omega})(1+A)]}$$

$$= \frac{R_i[(1+A)(1-\frac{\omega_o^2}{\omega^2})^2-A+\frac{\omega_o^2}{\omega^2}(1+A)+jQ_L\frac{\omega_o}{\omega}(1+A)^2(1-\frac{\omega_o^2}{\omega^2})+j\frac{1}{Q_L}(\frac{\omega}{\omega_o}\frac{A}{A+1}-\frac{\omega_o}{\omega})]}{1+Q_L^2\frac{\omega_o^2}{\omega^2}(1+A)^2}$$

$$= \frac{R_i\{1+j[Q_L\frac{\omega_o}{\omega}(1+A)^2(1-\frac{\omega_o^2}{\omega^2})+\frac{1}{Q_L}(\frac{\omega}{\omega_o}\frac{A}{A+1}-\frac{\omega_o}{\omega})]\}}{1+Q_L^2\frac{\omega_o^2}{\omega^2}(1+A)^2}.$$

Since $tan(Arg\mathbf{Z})$ is equal to the ratio of $Im(\mathbf{Z})$ over $Re(\mathbf{Z})$, the above equation yields

$$tan\psi = tan(Arg\mathbf{Z}) = \frac{1}{Q_L}\left(\frac{\omega}{\omega_o}\frac{A}{A+1}-\frac{\omega_o}{\omega}\right)$$

$$+Q_L\left(\frac{\omega_o}{\omega}\right)(1+A)^2\left[1-\left(\frac{\omega_o}{\omega}\right)^2\right].$$

9.3   The resonant circuit of the inverter of Fig. 9.1 has the following parameters: $L_1 = 300\ \mu H$, $L_2 = 200\ \mu H$, $C = 2$ nF, and $R_i = 600\ \Omega$. The circuit is driven by a sinusoidal voltage source $v = 100sin(1.41 \times 10^6)t$. What is the amplitude of the voltage across the ac load R_i in this circuit?

Equations (9.1), (9.2), (9.3), and (9.5) give the ratio of the inductances, the equivalent inductance, the corner frequency, and the loaded quality factor of the circuit, respectively,

$$A = \frac{L_1}{L_2} = \frac{300}{200} = 1.5$$

$$L = L_1 + L_2 = 300 + 200 = 500\ \mu H$$

$$\omega_o = \frac{1}{\sqrt{LC}} = \frac{1}{\sqrt{500 \times 10^{-6} \times 2 \times 10^{-9}}} = 1\frac{\text{Mrad}}{\text{s}}$$

$$Q_L = \omega_o C R_i = 10^6 \times 2 \times 10^{-9} \times 600 = 1.2.$$

The normalized switching frequency is $\omega/\omega_o = 1.41$. The voltage transfer function of the resonant circuit $\mathbf{M_{Vr}} \equiv V_{Ri}/V_{rms}$ can be also described $\mathbf{M_{Vr}} = V_{Rim}/V_m$. Thus, using (9.23),

$$V_{Rim} = V_m M_{Vr} = V_m\frac{1}{\sqrt{(1+A)^2[1-(\frac{\omega_o}{\omega})^2]^2+\frac{1}{Q_L^2}(\frac{\omega}{\omega_o}\frac{A}{A+1}-\frac{\omega_o}{\omega})^2}}$$

$$= 100 \times \frac{1}{\sqrt{(1+1.5)^2[1 - (\frac{1}{1.41})^2]^2 + \frac{1}{1.2^2}(1.41\frac{1.5}{1.5+1} - \frac{1}{1.41})^2}}$$

$$= 100 \times 0.8015 = 80.15 \text{ V}.$$

9.4 For the circuit of Problem 9.3, find the voltage stress across the resonant capacitor.

The ratio of the inductances, the loaded quality factor, the normalized switching frequency, and the voltage transfer function of the resonant circuit have been calculated in the solution to Problem 9.3. From (9.20) and (9.38), the maximum voltage across the resonant capacitor is

$$V_{Cm} = \left(\frac{\omega_o}{\omega}\right)\left(\frac{V_m M_{Vr}}{Q_L}\right)\sqrt{1 + \left[Q_L\left(\frac{\omega_o}{\omega}\right)(1+A)\right]^2}$$

$$= \left(\frac{1}{1.41}\right)\left(\frac{100 \times 0.8015}{1.2}\right)\sqrt{1 + \left[1.2\left(\frac{1}{1.41}\right)(1+1.5)\right]^2} = 111.4 \text{ V}.$$

9.5 Design a full-bridge CLL resonant inverter shown in Fig. 9.17 to meet the following specifications: $V_I = 250$ V and $P_{Rimax} = 85$ W. Assume the corner frequency $f_o = 100$ kHz, the normalized switching frequency $\omega/\omega_o = 1.5$, and the total efficiency of the inverter $\eta_I = 90\%$. Make the voltage transfer function of the inverter independent of the load.

It follows from (9.63) that the voltage transfer function of the inverter is independent of the load when the following condition must be satisfied

$$\frac{\omega}{\omega_o}\frac{A}{A+1} - \frac{\omega_o}{\omega} = 0.$$

Hence,

$$A = \frac{1}{\frac{\omega^2}{\omega_o^2} - 1} = \frac{1}{1.5^2 - 1} = 0.8.$$

Using (9.27) and (9.63), the voltage transfer function of the inverter is now

$$M_{VIa} = \frac{V_{Ri}}{V_I} = \frac{\sqrt{P_{Ri}R_i}}{V_I} = \frac{2\sqrt{2}\eta_I}{\pi(1+A)[1 - (\frac{\omega_o}{\omega})^2]}$$

from which

$$R_i = \frac{8\eta_I^2 V_I^2}{\pi^2 P_{Ri}(1+A)^2[1-(\frac{\omega_o}{\omega})^2]^2} = \frac{8 \times 0.9^2 \times 250^2}{\pi^2 \times 85(1+0.8)^2(1-\frac{1}{1.5^2})^2}$$

$$= 482.8 \ \Omega.$$

To assure good efficiency the loaded quality factor is selected according to (9.35)

$$Q_L = \frac{\frac{\omega}{\omega_o}}{1+A} = \frac{1.5}{1+0.8} = 0.833.$$

From (9.2) and (9.5), the values of the resonant components are

$$C = \frac{Q_L}{\omega_o R_i} = \frac{0.833}{2 \times \pi \times 100 \times 10^3 \times 482.8} = 2.75 \text{ nF}$$

$$L = \frac{R_i}{\omega_o Q_L} = \frac{482.8}{2 \times \pi \times 100 \times 10^3 \times 0.833} = 922.5 \ \mu\text{H}$$

$$L_1 = \frac{L}{1+\frac{1}{A}} = \frac{922.5}{1+\frac{1}{0.8}} = 410 \ \mu\text{H}$$

$$L_2 = \frac{L}{1+A} = \frac{922.5}{1+0.8} = 512.5 \ \mu\text{H}.$$

The characteristic impedance of the resonant circuit is

$$Z_o = \frac{R_i}{Q_L} = \frac{482.8}{0.833} = 579.6 \ \Omega.$$

From (9.65), the peak value of the current through the resonant circuit is

$$I_m = \frac{4V_I M_{Vr}}{\pi R_i}\sqrt{1 + \left[Q_L\left(\frac{\omega_o}{\omega}\right)(1+A)\right]^2}$$

$$= \frac{4\sqrt{2}V_I}{\pi R_i} = \frac{4\sqrt{2} \times 250}{\pi 482.8} = 0.932 \text{ A}.$$

The maximum value of the current through the switches is equal to I_m.

The peak values of the voltages across the reactive components can be calculated using (9.67) through (9.69):

$$V_{Cm} = \frac{I_m}{\omega C} = \frac{0.932}{1.5 \times 2 \times \pi \times 100 \times 10^3 \times 2.75 \times 10^{-9}} = 359.6 \text{ V}$$

$$V_{L1m} = \omega L_1 I_m = 1.5 \times 2 \times \pi \times 100 \times 10^3 \times 410 \times 10^{-6} \times 0.932$$

$$= 360.1 \text{ V}$$

and

$$V_{L2m} = \sqrt{2}V_{Ri} = \sqrt{2P_{Ri}R_i} = \sqrt{2 \times 85 \times 482.8} = 286.5 \text{ V}.$$

Chapter 10

10.1 A Class D series resonant inverter of Fig. 10.1 has the following parameters: $L = 220\ \mu\mathrm{H}$, $C = 3.3$ nF, $Q_L = 4$, and $f = 200$ kHz. Calculate the minimum dead time in the drive voltages that is required for achieving zero-voltage-switching operation if $C_1 = 500$ pF.

Using (10.9), one obtains the resonant frequency of the inverter

$$f_o = \frac{1}{2\pi\sqrt{LC}} = \frac{1}{2\pi\sqrt{220 \times 10^{-6} \times 3.3 \times 10^{-9}}} = 186.8\text{ kHz.}$$

Hence, the normalized switching frequency is

$$\frac{f}{f_o} = \frac{200}{186.8} = 1.071.$$

The characteristic impedance can be obtained from (10.10) as

$$Z_o = \sqrt{\frac{L}{C}} = \sqrt{\frac{220 \times 10^{-6}}{3.3 \times 10^{-9}}} = 258.2\ \Omega.$$

Using (10.22), the minimum dead time is calculated as

$$t_1 = \frac{\pi C_1 Z_o[\frac{1}{Q_L^2} + (\frac{f}{f_o} - \frac{f_o}{f})^2]}{2(\frac{f}{f_o} - \frac{f_o}{f})}$$

$$= \frac{\pi \times 500 \times 10^{-12} \times 258.2[\frac{1}{4^2} + (1.071 - \frac{1}{1.071})^2]}{2(1.071 - \frac{1}{1.071})} = 0.12\ \mu\text{s.}$$

10.2 The circuit of Problem 10.1 has sinusoidal drive voltages of power MOSFETs. Select the amplitude of the drive signals which assures that shunt capacitor C_1 is charged and discharged completely. Assume that MOSFETs with a threshold voltage $V_T = 2$ V are used.

Using (10.29), the maximum amplitude of the sinusoidal drive voltage can be expressed as

$$V_{GSm} = \frac{V_T}{sin(\pi f t_1)} = \frac{2}{sin(\pi \times 200 \times 10^3 \times 0.12 \times 10^{-6})} = 26.55\text{ V.}$$

10.3 Derive equation (10.25).

From (10.4) and (10.7), the charging time is

$$t_1 = \frac{C_1 V_I}{I}$$

where

$$I = I_m sin\psi.$$

For the parallel resonant inverter, the amplitude of the current through the resonant inductor is given by (7.48)

$$I_m = \frac{2V_I}{\pi Z_o} \sqrt{\frac{1 + (Q_L \frac{\omega}{\omega_o})^2}{Q_L^2 [1 - (\frac{\omega}{\omega_o})^2]^2 + (\frac{\omega}{\omega_o})^2}}.$$

From trigonometric relationships and (7.13),

$$sin\psi = \frac{Q_L(\frac{\omega}{\omega_o})[(\frac{\omega}{\omega_o})^2 + \frac{1}{Q_L^2} - 1]}{\sqrt{1 + \{Q_L(\frac{\omega}{\omega_o})[(\frac{\omega}{\omega_o})^2 + \frac{1}{Q_L^2} - 1]\}^2}}.$$

Hence, the charging time is

$$t_1 = \frac{\pi C_1 Z_o}{2} \sqrt{\frac{[1 - (\frac{\omega}{\omega_o})^2]^2 + (\frac{1}{Q_L}\frac{\omega}{\omega_o})^2}{1 + (Q_L \frac{\omega}{\omega_o})^2}}$$

$$\times \frac{\sqrt{1 + \{Q_L(\frac{\omega}{\omega_o})[(\frac{\omega}{\omega_o})^2 + \frac{1}{Q_L^2} - 1]\}^2}}{(\frac{\omega}{\omega_o})[(\frac{\omega}{\omega_o})^2 + \frac{1}{Q_L^2} - 1]}.$$

Chapter 11

11.1 In the parallel resonant circuit of Fig. 11.1, the load resistance is $R_i = 500\ \Omega$ and the efficiency is $\eta_{rc} = 99\%$. What is the EPR of the resonant inductor if the EPR of the resonant capacitor is $R_{Cp} = 100\ \text{k}\Omega$?

From (11.4),

$$R_{Lp} = \frac{R_d R_{Cp}}{R_{Cp} - R_d}$$

and, from (11.14),

$$R_d = \frac{R_i \eta_{rc}}{1 - \eta_{rc}}.$$

Hence,

$$R_{Lp} = \frac{\frac{R_i \eta_{rc}}{1-\eta_{rc}} R_{Cp}}{R_{Cp} - \frac{R_i \eta_{rc}}{1-\eta_{rc}}} = \frac{\frac{500 \times 0.99}{1-0.99} \times 100 \times 10^3}{100 \times 10^3 - \frac{500 \times 0.99}{1-0.99}} = 98.02\ \text{k}\Omega.$$

11.2 A parallel resonant circuit with $R = 300\ \Omega$, $L = 500\ \mu\text{H}$, and $C = 2\ \text{nF}$ is driven by a sinusoidal current source $i = 2sin(800 \times 10^3 t)$. Calculate the power dissipate in this circuit.

The resonant frequency is calculated from (11.7) as

$$\omega_o = \frac{1}{\sqrt{LC}} = \frac{1}{\sqrt{500 \times 10^{-6} \times 2 \times 10^{-9}}} = 10^6\ \frac{\text{rad}}{\text{s}}$$

and (11.8) gives the characteristic impedance of the resonant circuit

$$Z_o = \sqrt{\frac{L}{C}} = \sqrt{\frac{500 \times 10^{-6}}{2 \times 10^{-9}}} = 500\ \Omega.$$

The normalized switching frequency is

$$\frac{\omega}{\omega_o} = \frac{800 \times 10^3}{10^6} = 0.8.$$

The power dissipated in the resonant circuit is given by

$$P_{Rd} = \frac{V_{Ri}^2}{R_d}.$$

Using (11.12) and (11.9), the rms value of the voltage across the resonant circuit is

$$V_{Ri} = \frac{I_{rms}}{Y} = \frac{I_{rms}}{\sqrt{\frac{1}{R^2} + \frac{1}{Z_o^2}\left(\frac{\omega}{\omega_o} - \frac{\omega_o}{\omega}\right)^2}}.$$

Thus,

$$P_R = \frac{I_{rms}^2 R}{1 + \frac{R^2}{Z_o^2}\left(\frac{\omega}{\omega_o} - \frac{\omega_o}{\omega}\right)^2} = \frac{(\frac{2}{\sqrt{2}})^2 \times 300}{1 + \frac{300^2}{500^2}(0.8 - \frac{1}{0.8})^2} = 559.2 \text{ W}.$$

11.3 Assume that the frequency of the current source of Problem 11.2 can be adjusted. At what frequency are the current and voltage stresses of the resonant components maximum? Calculate those stresses.

The current stresses of the resonant components have maximum values at the resonant frequency $\omega_o = 10^6$ rad/s (see Problem 11.2) and, using (11.9) and (11.33), are obtained as

$$I_{Lm} = I_{Cm} = Q_L I_m = R I_m \sqrt{\frac{C}{L}} = 300 \times 2\sqrt{\frac{2 \times 10^{-9}}{500 \times 10^{-6}}} = 1.2 \text{ A}.$$

The voltage stress has also the maximum value at resonant frequency because the whole input current flows through the resistance R.

$$V_{Rim} = V_{Lm} = V_{Cm} = R I_m = 300 \times 2 = 600 \text{ V}.$$

11.4 Derive equation (11.25).

Rearrangement of (11.24) gives

$$\frac{2M_{VI}^2 \eta_{rc}^2}{\pi^2 \eta_I^2} = 1 + \left[Q_L\left(\frac{\omega}{\omega_o} - \frac{\omega_o}{\omega}\right)\right]^2$$

$$\frac{2M_{VI}^2 \eta_{rc}^2}{\pi^2 \eta_I^2 Q_L^2} - \frac{1}{Q_L^2} = \left(\frac{\omega}{\omega_o} - \frac{\omega_o}{\omega}\right)^2$$

$$\sqrt{a} = \left|\frac{\omega}{\omega_o} - \frac{\omega_o}{\omega}\right|$$

where

$$a = \frac{2M_{VI}^2 \eta_{rc}^2}{\pi^2 \eta_I^2 Q_L^2} - \frac{1}{Q_L^2} > 0.$$

For $\omega/\omega_o = f/f_o < 1$, the above equations lead to a quadratic equation with respect to f/f_o of the form

$$\frac{f^2}{f_o^2} + \frac{f}{f_o}\sqrt{a} - 1 = 0.$$

The solution of this quadratic equation gives (11.25)

$$\frac{f}{f_o} = \frac{-\sqrt{a} + \sqrt{a+4}}{2}.$$

11.5 Design a current–source inverter of Fig. 11.1. The following specifications should be satisfied: $V_I = 200$ V, $R_{imin} = 1000$ Ω, and $P_{Rimax} = 200$ W. Assume the resonant frequency $f_o = 100$ kHz, the normalized switching frequency $f/f_o = 0.95$, the total efficiency of the inverter $\eta_I = 95\%$, and the ratio $R/R_i = 0.99$.

The total resistance of the inverter is $R = R_i(R/R_i) = 1000 \times 0.99 = 990$ Ω. From (11.24) and the relationship $V_{Ri} = \sqrt{P_{Ri}R_i}$, the loaded quality factor is

$$Q_L = \frac{\sqrt{\frac{2R^2 P_{Ri}}{\pi^2 \eta_I^2 R_i V_I^2} - 1}}{\left| \frac{\omega}{\omega_o} - \frac{\omega_o}{\omega} \right|} = \frac{\sqrt{\frac{2 \times 990^2 \times 200}{\pi^2 \times 0.95^2 \times 1000 \times 200^2} - 1}}{\left| 0.95 - \frac{1}{0.95} \right|} = 3.09.$$

Using (11.9), the values of the resonant components can be calculated as

$$L = \frac{R}{\omega_o Q_L} = \frac{990}{2 \times \pi \times 100 \times 10^3 \times 3.09} = 510 \ \mu\text{H}$$

and

$$C = \frac{Q_L}{\omega_o R} = \frac{3.09}{2 \times \pi \times 100 \times 10^3 \times 990} = 4.97 \ \text{nF}.$$

The maximum dc input power is

$$P_{Imax} = \frac{P_{Rimax}}{\eta_I} = \frac{200}{0.95} = 210.5 \ \text{W}$$

and the maximum value of the dc input current equal to the maximum value of the switch current is

$$I_{I(max)} = I_{SM(max)} = \frac{P_{Imax}}{V_I} = \frac{210.5}{200} = 1.05 \ \text{A}.$$

The maximum value of the switch voltage is

$$V_{SM} = \sqrt{2P_{Rimax}R_{imin}} = \sqrt{2 \times 200 \times 1000} = 632.5 \ \text{V}.$$

Chapter 12

12.1 A single-capacitor phase-controlled series resonant inverter operates at a normalized switching frequency $f/f_o = 1.25$. The phase shift at the full load is $\phi = 20°$ and the maximum value of the loaded quality factor is $Q_L = 3$. What is the phase shift at 50% of the full load?

From (12.14), the voltage transfer function of the inverter at full load is

$$M_{VI} = \frac{\sqrt{2}cos(\frac{\phi}{2})}{\pi\sqrt{1 + Q_L^2(\frac{\omega}{\omega_o} - \frac{\omega_o}{\omega})^2}} = \frac{\sqrt{2}cos(\frac{20°}{2})}{\pi\sqrt{1 + 3^2(1.25 - \frac{1}{1.25})^2}} = 0.2639.$$

It follows from the definition of Q_L that at 50% of the full load, the loaded quality factor is half of that for the minimum load resistance, i.e., $Q_L = 1.5$. Using (12.14), the phase shift at 50% of full load can be expressed as

$$\phi = 2arccos\left[\frac{\pi M_{VI}\sqrt{1 + Q_L^2(\frac{\omega}{\omega_o} - \frac{\omega_o}{\omega})^2}}{\sqrt{2}}\right]$$

$$= 2arccos\left[\pi \times \frac{0.2639\sqrt{1 + 1.5^2(1.25 - \frac{1}{1.25})^2}}{\sqrt{2}}\right] = 89.97°.$$

12.2 A single-capacitor phase-controlled series resonant inverter operating at a switching frequency $f = 200$ kHz has the following parameters: $V_I = 180$ V, $L = 400$ μH, $C = 4.7$ nF, $\phi = 25°$, and $R_i = 50$ Ω. Calculate the output power of the inverter.

The resonant frequency of the inverter is

$$f_o = \frac{1}{\pi\sqrt{2LC}} = \frac{1}{\pi\sqrt{2 \times 400 \times 10^{-6} \times 4.7 \times 10^{-9}}} = 164.2 \text{ kHz}$$

and the loaded quality factor is

$$Q_L = \frac{Z_o}{2R_i} = \frac{\sqrt{\frac{2L}{C}}}{2R_i} = \frac{\sqrt{\frac{2 \times 400 \times 10^{-6}}{4.7 \times 10^{-9}}}}{2 \times 50} = 4.126.$$

From (12.25),

$$P_{Ri} = \frac{2V_I^2 cos^2(\frac{\phi}{2})}{\pi^2 R_i[1 + Q_L^2(\frac{\omega}{\omega_o} - \frac{\omega_o}{\omega})^2]}$$

$$= \frac{2 \times 180^2 cos^2(\frac{25}{2})}{\pi^2 \times 50[1 + 4.126^2(\frac{200}{164.2} - \frac{164.2}{200})^2]} = 33.98 \text{ W}.$$

12.3 An equivalent circuit for the fundamental component of the phase-controlled Class D series resonant inverter [18] is depicted in Fig. 12.6, where $v_1 = V_m cos(\omega t + \phi/2)$ and $v_2 = V_m cos(\omega t - \phi/2)$. Find the voltage transfer function $M_{VI} = V_{Ri}/V_m$ of the inverter in terms of phase shift ϕ, normalized switching frequency ω/ω_o, where $\omega_o = 1/\sqrt{LC}$, and loaded quality factor $Q_L = \omega_o L/(2R_i)$. Compare your result to (12.13).

The voltages v_1 and v_2 can be expressed in the complex domain by

$$\mathbf{V_1} = V_m e^{j(\phi/2)}$$

and

$$\mathbf{V_2} = V_m e^{-j(\phi/2)}.$$

Let us apply the principle of superposition. The output voltages caused by the voltages $\mathbf{V_1}$ and $\mathbf{V_2}$ are:

$$\mathbf{V_{o1}} = \frac{R_i V_m e^{j(\phi/2)}}{2R_i + jX}$$

and

$$\mathbf{V_{o2}} = \frac{R_i V_m e^{-j(\phi/2)}}{2R_i + jX}$$

where

$$X = \omega L - \frac{1}{\omega C} = \sqrt{\frac{L}{C}}\left(\frac{\omega}{\omega_o} - \frac{\omega_o}{\omega}\right) = 2R_i Q_L \left(\frac{\omega}{\omega_o} - \frac{\omega_o}{\omega}\right).$$

Hence, the output voltage is

$$\mathbf{V_o} = \mathbf{V_{o1}} + \mathbf{V_{o2}} = \frac{V_m cos(\frac{\phi}{2})}{1 + jQ_L(\frac{\omega}{\omega_o} - \frac{\omega_o}{\omega})}.$$

Rearrangement of the above equation gives the voltage transfer function as

$$\mathbf{M_{VI}} = \frac{\mathbf{V_o}}{\sqrt{2}V_m} = \frac{cos(\frac{\phi}{2})}{\sqrt{2}[1 + jQ_L(\frac{\omega}{\omega_o} - \frac{\omega_o}{\omega})]}.$$

Taking into account that $V_m = 2V_I/\pi$, this voltage transfer function turns out to be the same as the voltage transfer function of the single-capacitor PC SRC given by (12.13). However, at the same output power, the currents through the resonant inductors in the single-capacitor PC SRC are smaller which results in a better efficiency of this converter.

12.4 Fig. 12.7 shows an equivalent circuit for the fundamental component of the phase-controlled Class D parallel resonant inverter [22], where $v_1 = V_m cos(\omega t + \phi/2)$ and $v_2 = V_m cos(\omega t - \phi/2)$. Derive an expression for the voltage transfer function $M_{VI} = V_{Ri}/V_m$ of the inverter in terms of the phase shift ϕ, normalized switching frequency ω/ω_o, where $\omega_o = \sqrt{2/LC}$, and loaded quality factor $Q_L = 2R_i/(\omega_o L)$.

The phasors of the voltages at the input of the resonant circuits are expressed by

$$\mathbf{V_1} = V_m e^{j(\phi/2)}$$

and

$$\mathbf{V_2} = V_m e^{-j(\phi/2)}.$$

The voltages across the parallel connection C–R_i caused by voltage sources $\mathbf{V_1}$ and $\mathbf{V_2}$ separately, i.e., with the other voltage source shorted, can be expressed as

$$\mathbf{V_{o1}} = \frac{\mathbf{V_1} R_i \parallel \frac{1}{j\omega C} \parallel j\omega L}{j\omega L + R_i \parallel \frac{1}{j\omega C} \parallel j\omega L} = \frac{V_m e^{j(\phi/2)}}{2[1 - (\frac{\omega}{\omega_o})^2 + j\frac{1}{Q_L}\frac{\omega}{\omega_o}]}$$

and

$$\mathbf{V_{o2}} = \frac{\mathbf{V_2} R_i \parallel \frac{1}{j\omega C} \parallel j\omega L}{j\omega L + R_i \parallel \frac{1}{j\omega C} \parallel j\omega L} = \frac{V_m e^{-j(\phi/2)}}{2[1 - (\frac{\omega}{\omega_o})^2 + j\frac{1}{Q_L}\frac{\omega}{\omega_o}]}.$$

Using the principle of superposition,

$$\mathbf{V_o} = \mathbf{V_{o1}} + \mathbf{V_{o2}} = \frac{V_m cos(\frac{\phi}{2})}{1 - (\frac{\omega}{\omega_o})^2 + j\frac{1}{Q_L}\frac{\omega}{\omega_o}}.$$

Hence, the voltage transfer function of the phase-controlled parallel resonant inverter is

$$\mathbf{M_{VI}} = \frac{\mathbf{V_o}}{\sqrt{2}V_m} = \frac{cos(\frac{\phi}{2})}{\sqrt{2}[1 - (\frac{\omega}{\omega_o})^2 + j\frac{1}{Q_L}\frac{\omega}{\omega_o}]}.$$

12.5 Design a single-capacitor phase-controlled series resonant inverter of Fig. 12.1 that delivers 100 W power to 25 Ω load resistance. The input voltage of the inverter is $V_I = 180$ V. Assume the resonant frequency $f_o = 100$ kHz, the normalized switching frequency $f/f_o = 1.25$, $cos(\phi/2) = 0.9$, and the inverter efficiency $\eta_I = 94\%$.

The rms value of the output voltage can be found as

$$V_{o(rms)} = \sqrt{P_{Ri} R_i} = \sqrt{100 \times 25} = 50 \text{ V}.$$

Using (12.14) and (12.15), the quality factor at full load is

$$Q_L = \frac{1}{\left|\frac{\omega}{\omega_o} - \frac{\omega_o}{\omega}\right|}\sqrt{\frac{2\eta_I^2 V_I^2 cos^2(\frac{\phi}{2})}{V_{o(rms)}^2 \pi^2} - 1}$$

$$= \frac{1}{\left|1.25 - \frac{1}{1.25}\right|}\sqrt{\frac{2 \times 0.94^2 \times 180^2 \times 0.9^2}{50^2 \pi^2} - 1} = 2.08.$$

Thus,

$$L = \frac{2R_i Q_L}{\omega_o} = \frac{2 \times 25 \times 2.08}{2\pi \times 100 \times 10^3} = 165.5 \text{ } \mu\text{H}$$

and

$$C = \frac{2}{\omega_o^2 L} = \frac{2}{(2\pi \times 100 \times 10^3)^2 \times 165.5 \times 10^{-6}} = 30.6 \text{ nF}.$$

Chapter 13

13.1 Design an optimum Class E ZVS inverter to meet the following specifications: P_{Ri} = 125 W, V_I = 48 V, and f = 2 MHz. Assume $Q_L = 5$.

Let us assume $D = 0.5$. From (13.48), the full-load resistance is

$$R_i = \frac{8}{\pi^2 + 4} \frac{V_I^2}{P_{Ri}} = 0.5768 \times \frac{48^2}{125} = 10.63 \ \Omega.$$

The dc input current can be obtained using (13.41)

$$I_I = \frac{8}{\pi^2 + 4} \frac{V_I}{R_i} = 0.5768 \times \frac{48}{10.63} = 2.6 \text{ A}.$$

Hence, from (13.42), the maximum switch current is

$$I_{SM} = \left(\frac{\sqrt{\pi^2 + 4}}{2} + 1 \right) I_I = 2.862 \times 2.6 = 7.44 \text{ A}$$

and, from (13.45), the maximum amplitude of the current through the resonant circuit is

$$I_m = \frac{\sqrt{\pi^2 + 4}}{2} I_I = 1.8621 \times 2.6 = 4.84 \text{ A}.$$

The maximum switch voltage is given by (13.43)

$$V_{SM} = 3.652 V_I = 3.652 \times 48 = 175.3 \text{ V}.$$

Using (13.31), (13.49), and (13.52), the values of the load–network components are:

$$L = \frac{Q_L R_i}{\omega} = \frac{5 \times 10.63}{2\pi \times 2 \times 10^6} = 4.23 \ \mu\text{H}$$

$$C_1 = \frac{4}{\pi^2(\pi^2 + 4) f R_i} = \frac{4}{\pi^2(\pi^2 + 4) \times 2 \times 10^6 \times 10.63} = 1.375 \text{ nF}$$

and

$$C = \frac{1}{\omega R_i \left[Q_L - \frac{\pi(\pi^2 - 4)}{16} \right]}$$

$$= \frac{1}{2\pi \times 2 \times 10^6 \times 10.63(5 - 1.1525)} = 1.95 \text{ nF}.$$

From (13.33), the minimum value of the choke inductance is

$$L_f = 2\left(\frac{\pi^2}{4}+1\right)\frac{R_i}{f} = 2\left(\frac{\pi^2}{4}+1\right)\times\frac{10.63}{2\times10^6} = 36.9~\mu\text{H}.$$

13.2 The rms value of the US utility voltage is from 92 to 132 V. This voltage is rectified by a bridge peak rectifier to supply a Class E ZVS inverter that is operated at a switch duty cycle of 0.5. What is the required value of the voltage rating of the switch?

The maximum value of the input voltage is

$$V_{Imax} = \sqrt{2}V_{rms(max)} = \sqrt{2}\times132 = 186.7~\text{V}.$$

From (13.43), the maximum switch voltage is

$$V_{SM} = 3.652V_{Imax} = 3.652\times186.7 = 682~\text{V}.$$

Considering a safety factor, a switch with a voltage rating of at least 800 V should be used.

13.3 Repeat Problem 13.2 for the European utility line whose rms voltage is $220\pm15\%$.

The maximum input voltage is in this case

$$V_{Imax} = \sqrt{2}(V_{rms}+0.15V_{rms}) = \sqrt{2}\times1.15\times220 = 357.8~\text{V}.$$

Using (13.43), the maximum switch voltage is

$$V_{SM} = 3.652V_{Imax} = 3.652\times357.8 = 1307~\text{V}.$$

The voltage rating of a switch should be about 1500 V. Power MOSFETs are presently well below this value and a designer should consider an application of very high power switches like thyristors, IGBTs, or MCTs.

13.4 Derive the design equations for the component values for the matching resonant circuit $\pi2a$ shown in Fig. 13.12(b).

An equivalent circuit of the matching circuit of Fig. 13.13(b) is shown in Fig. P13.1.

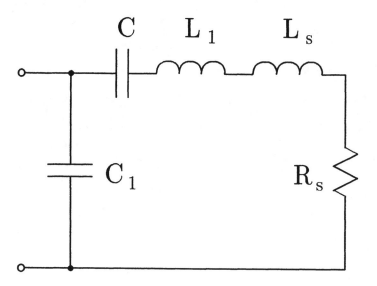

Figure P13.1: An equivalent circuit of the matching circuit of Fig. 13.11(b).

The values of X_{C1}, X_C, and R_s can be computed from (13.71), (13.73), and (13.70), respectively. The reactance factor of the R_i–L_2 and R_s–L_s two–port networks is

$$q = \frac{R_i}{X_{L2}} = \frac{X_{Ls}}{R_s}.$$

The series equivalent components are

$$R_s = \frac{R_i}{1 + q^2} = \frac{R_i}{1 + \left(\frac{R_i}{X_{L2}}\right)^2}$$

and

$$X_{Ls} = \frac{X_{L2}}{1 + \frac{1}{q^2}}.$$

Hence,

$$q = \sqrt{\frac{R_i}{R_s} - 1}$$

and

$$X_{Ls} = qR_s = R_s\sqrt{\frac{R_i}{R_s} - 1}.$$

Using the definition of the loaded quality factor,

$$X_L = \omega(L_1 + L_s) = Q_L R_s.$$

Thus,

$$X_{L1} = \omega L_1 = X_L - X_{Ls} = (Q_L - q)R_s = \left(Q_L - \sqrt{\frac{R_i}{R_s} - 1} \right) R_s.$$

Using both equations for q

$$X_{L2} = \frac{R_i}{q} = \frac{R_i}{\sqrt{\frac{R_i}{R_s} - 1}}.$$

13.5 Find the maximum operating frequency at which pure Class E operation is still achievable for $V_I = 200$ V, $P_{Ri} = 75$ W, and $C_{out} = 100$ pF.

Using (13.69), the maximum frequency is

$$f_{max} = \frac{0.05439 P_{Ri}}{V_I^2 C_{out}} = \frac{0.05439 \times 75}{200^2 \times 100 \times 10^{-12}} = 1.069 \text{ MHz}.$$

Chapter 14

14.1 A Class E ZCS inverter is powered from a 340-V power supply. What is the required voltage rating of the switch if the switch duty cycle is 0.5?

From (14.31), the maximum voltage across the switch is

$$V_{SM} = \left(\frac{\sqrt{\pi^2 + 4}}{2} + 1\right) V_I = 2.8621 \times 340 = 973.1 \text{ V}.$$

14.2 Design a Class E ZCS inverter to meet the following specifications: $V_I = 180$ V, $P_{Ri} = 250$ W, and $f = 200$ kHz.

The full-load resistance of the inverter can be calculated from (14.56)

$$R_i = \frac{8}{\pi^2(\pi^2 + 4)} \frac{V_I^2}{P_{Ri}} = 0.05844 \times \frac{180^2}{250} = 7.57 \ \Omega.$$

Assuming $Q_L = 5$ and using (14.57), (14.59), and (14.60), the values of the reactive components can be calculated as

$$L_1 = \frac{\pi^2 + 4}{16} \frac{R_i}{f} = 0.8669 \times \frac{7.57}{200 \times 10^3} = 32.8 \ \mu\text{H}$$

$$C = \frac{1}{\omega R_i Q_L} = \frac{1}{2 \times \pi \times 200 \times 10^3 \times 7.57 \times 5} = 21 \text{ nF}$$

and

$$L = \left[Q_L - \frac{\pi(\pi^2 + 12)}{16}\right] \frac{R_i}{\omega}$$

$$= \left[5 - \frac{\pi(\pi^2 + 12)}{16}\right] \times \frac{7.57}{2 \times \pi \times 200 \times 10^3} = 4.25 \ \mu\text{H}.$$

Using (14.54), the dc input current is

$$I_I = \frac{8}{\pi^2(\pi^2 + 4)} \frac{V_I}{R_i} = 0.0584 \times \frac{180}{7.57} = 1.39 \text{ A}.$$

The maximum switch current is calculated from (14.30) as

$$I_{SM} = \pi(\pi - 2\varphi)I_I = 3.562 \times 1.39 = 4.95 \text{ A}$$

and (14.53) gives the maximum amplitude of the current through the resonant circuit

$$I_m = \frac{4}{\pi\sqrt{\pi^2+4}}\frac{V_I}{R_i} = 0.3419 \times \frac{180}{7.57} = 8.13 \text{ A}.$$

Using (14.31), the maximum voltage across the switch is

$$V_{SM} = \left(\frac{\sqrt{\pi^2+4}}{2}+1\right)V_I = 2.8621 \times 180 = 515.2 \text{ V}.$$

14.3 It has been found that a Class E ZCS inverter has the following parameters: $D = 0.5$, $f = 400$ kHz, $L_1 = 20$ μH, and $P_{Ri} = 100$ W. What is the maximum voltage across the switch in this inverter?

From (14.57), the load resistance is

$$R_i = \frac{16fL_1}{\pi^2+4} = \frac{16 \times 400 \times 10^3 \times 20 \times 10-6}{\pi^2+4} = 9.23 \text{ } \Omega.$$

Rearrangement of (14.56) yields the input voltage

$$V_I = \sqrt{\frac{\pi^2(\pi^2+4)P_{Ri}R_i}{8}} = \sqrt{17.11 \times 100 \times 9.23} = 125.7 \text{ V}.$$

Using (14.31), the maximum voltage across the switch can be obtained as

$$V_{SM} = \left(\frac{\sqrt{\pi^2+4}}{2}+1\right)V_I = 2.8621 \times 125.7 = 359.7 \text{ V}.$$

Chapter 15

15.1 A full–bridge series resonant converter with a transformer center–tapped rectifier has the following parameters: input voltage $V_I = 200$ V, output voltage $V_O = 24$ V, full-load resistance $R_L = 5$ Ω, inverter efficiency $\eta_I = 94\%$, rectifier efficiency $\eta_R = 90\%$, resonant inductance $L = 400$ μH, resonant capacitance $C = 1.5$ nF, and transformer turns ratio $n = 5$. Find the full-load operating frequency above the resonant frequency for this converter.

From (2.72), the input resistance of the rectifier is

$$R_i = \frac{8n^2 R_L}{\pi^2 \eta_R} = \frac{8 \times 5^2 \times 5}{\pi^2 \times 0.9} = 112.6 \ \Omega.$$

Using (6.69), the overall resistance of the inverter is

$$R = \frac{R_i}{\eta_I} = \frac{112.6}{0.94} = 119.8 \ \Omega.$$

The loaded quality factor of the inverter is given by (6.10)

$$Q_L = \frac{\sqrt{\frac{L}{C}}}{R} = \frac{\sqrt{\frac{400 \times 10^{-6}}{1.5 \times 10^{-9}}}}{119.8} = 4.31.$$

Using (2.73), the voltage transfer function of the rectifier is

$$M_{VR} = \frac{\pi \eta_R}{2\sqrt{2}n} = \frac{\pi \times 0.9}{2\sqrt{2} \times 5} = 0.2.$$

From (15.6), the voltage transfer function of the inverter is

$$M_{VI} = \frac{M_V}{M_{VR}} = \frac{V_O}{V_I M_{VR}} = \frac{24}{200 \times 0.2} = 0.6.$$

Rearrangement of (6.166) gives

$$\left| \frac{\omega}{\omega_o} - \frac{\omega_o}{\omega} \right| = \frac{\sqrt{\frac{8\eta_I^2}{\pi^2 M_{VI}^2} - 1}}{Q_L} = \frac{\sqrt{\frac{8 \times 0.94^2}{\pi^2 \times 0.6^2} - 1}}{4.31} = 0.2308$$

and, finally,

$$\frac{\omega}{\omega_o} = 1.122.$$

Hence, the full-load operating frequency of the converter is

$$f = \frac{\frac{\omega}{\omega_o}}{2\pi\sqrt{LC}} = \frac{1.122}{2\pi\sqrt{400 \times 10^{-6} \times 1.5 \times 10^{-9}}} = 230.5 \text{ kHz}.$$

15.2 A half–wave rectifier with a diode threshold voltage $V_F = 0.4$ V, diode forward resistance $R_F = 0.025$ Ω, ESR of the filter capacitor $r_C = 20$ mΩ, transformer turns ratio $n = 5$, and efficiency of the transformer $\eta_{tr} = 96\%$ is used in a half–bridge series resonant converter. The efficiency of the inverter is $\eta_I = 92\%$. The converter output voltage is $V_O = 5$ V and the maximum output current is $I_O = 10$ A. Calculate the overall efficiency of the converter and its input voltage if $Q_L = 4$ and $f/f_o = 1.1$ at full load. Neglect switching losses and the drive power.

The full-load resistance of the converter is

$$R_L = \frac{V_O}{I_O} = \frac{5}{10} = 0.5 \ \Omega.$$

Using (2.27), the efficiency of the rectifier can be found as

$$\eta_R = \frac{\eta_{tr}}{1 + \frac{2V_F}{V_O} + \frac{\pi^2 R_F}{2R_L} + \frac{r_C}{R_L}(\frac{\pi^2}{4} - 1)}$$

$$= \frac{0.96}{1 + \frac{2\times0.4}{5} + \frac{\pi^2\times0.025}{2\times0.5} + \frac{0.02}{0.5}(\frac{\pi^2}{4} - 1)} = 65.5\%.$$

Hence, from (15.3), the overall efficiency of the converter is

$$\eta = \eta_I\eta_R = 0.92 \times 0.655 = 60.3\%.$$

The voltage transfer function of the rectifier is given by (2.31)

$$M_{VR} = \frac{\pi\eta_R}{n\sqrt{2}} = \frac{\pi \times 0.655}{5\sqrt{2}} = 0.291.$$

From (15.4) and (6.64), the input voltage of the converter is

$$V_I = \frac{V_O}{M_{VI}M_{VR}} = \frac{V_O\pi\sqrt{1 + Q_L^2(\frac{\omega}{\omega_o} - \frac{\omega_o}{\omega})^2}}{\sqrt{2}\eta_I M_{VR}}$$

$$= \frac{5\pi\sqrt{1 + 4^2(1.1 - \frac{1}{1.1})^2}}{\sqrt{2} \times 0.92 \times 0.291} = 52.2 \text{ V}.$$

15.3 Design a half–bridge SRC with a bridge rectifier to meet the following specifications: input voltage $V_I = 110$ V, output voltage $V_O = 270$ V, and minimum load resistance $R_L = 500$ Ω. The parameters of the rectifier are: $V_F = 0.7$ V, $R_F = 0.1$ Ω, and $r_C = 25$ mΩ. Assume that the total inverter efficiency is $\eta_I = 0.9$, the resonant frequency is $f_o = 200$ kHz, and the switching frequency at full load is $f = 208$ kHz.

It can be seen that we are to design a step-up converter. Since the inverter and the transformerless version of the rectifier result in a step-down converter, the step-up conversion should be achieved by means of a transformer. Let us assume the transformer turns ratio $n = 1/6$ and the transformer efficiency $\eta_{tr} = 96\%$. Using (2.96), the rectifier efficiency is obtained as

$$\eta_R = \frac{\eta_{tr}}{1 + \frac{2V_F}{V_O} + \frac{\pi^2 R_F}{4R_L} + \frac{r_C}{R_L}\left(\frac{\pi^2}{8} - 1\right)}$$

$$= \frac{0.96}{1 + \frac{2\times0.7}{270} + \frac{\pi^2\times0.1}{4\times500} + \frac{0.025}{500}\left(\frac{\pi^2}{8} - 1\right)} = 95.46\%.$$

The input resistance of the rectifier is given by (2.97)

$$R_i = \frac{8n^2 R_L}{\pi^2 \eta_R} = \frac{\frac{8}{6^2}\times500}{\pi^2\times0.9546} = 11.8\ \Omega.$$

From (2.98), the voltage transfer function of the rectifier is

$$M_{VR} = \frac{\pi \eta_R}{2\sqrt{2}n} = \frac{\pi\times0.9546}{2\sqrt{2}\times\frac{1}{6}} = 6.362.$$

From (15.6), the voltage transfer function of the inverter is

$$M_{VI} = \frac{M_V}{M_{VR}} = \frac{V_O}{V_I M_{VR}} = \frac{270}{110\times6.362} = 0.3858.$$

Rearrangement of (6.64) gives

$$Q_L = \frac{\sqrt{\frac{2\eta_I^2}{\pi^2 M_{VI}^2} - 1}}{\left|\frac{\omega}{\omega_o} - \frac{\omega_o}{\omega}\right|} = \frac{\sqrt{\frac{2\times0.9^2}{\pi^2\times0.3858^2} - 1}}{\frac{208}{200} - \frac{200}{208}} = 4.09.$$

The overall resistance of the inverter is

$$R = \frac{R_i}{\eta_I} = \frac{11.8}{0.9} = 13.1\ \Omega.$$

Hence, using (6.10), the values of the resonant components are

$$L = \frac{Q_L R}{\omega_o} = \frac{4.09 \times 13.1}{2 \times \pi \times 200 \times 10^3} = 42.6 \ \mu\text{H}$$

$$C = \frac{1}{\omega_o Q_L R} = \frac{1}{2 \times \pi \times 200 \times 10^3 \times 4.09 \times 13.1} = 14.85 \ \text{nF}.$$

The voltage stresses of the resonant components are obtained from (6.49) as

$$V_{Cm} = V_{Lm} = \frac{2V_I Q_L}{\pi} = \frac{2 \times 110 \times 4.09}{\pi} = 286.4 \ \text{V}.$$

Using (6.36), the maximum switch current at full power is

$$I_{SM(max)} = I_{m(max)} = \frac{2V_I}{\pi R \sqrt{1 + Q_L^2(\frac{\omega}{\omega_o} - \frac{\omega_o}{\omega})^2}}$$

$$= \frac{2 \times 110}{\pi \times 13.2\sqrt{1 + 4.09^2(\frac{208}{200} - \frac{200}{208})^2}} = 5.1 \ \text{A}.$$

The maximum switch voltage is equal to the input voltage. From (2.91) and (2.92), the stresses of the rectifier diodes are

$$I_{DM} = \frac{\pi I_O}{2} = \frac{\pi V_O}{2R_L} = \frac{\pi \times 270}{2 \times 500} = 0.85 \ \text{A}$$

and

$$V_{DM} = V_O = 270 \ \text{V}.$$

Chapter 16

16.1 A full-bridge PRC with a bridge rectifier has the following parameters: $M_V = 5$, $\eta_I = 95\%$, $\eta_R = 92\%$, $n = 1/3$, and $\omega/\omega_o = 0.9$. The input resistance of the rectifier is $R_i = 500$ Ω. What is the value of the resonant inductance L if the resonant capacitance is $C = 4.7$ nF?

From (16.26), the voltage transfer function of the converter is

$$M_V = \frac{8\eta_I\eta_R}{n\pi^2\sqrt{[1 - (\frac{\omega}{\omega_o})^2]^2 + [\frac{1}{Q_L}(\frac{\omega}{\omega_o})]^2}}.$$

Rearrangement of the above equation gives the loaded quality factor as

$$Q_L = \frac{\frac{\omega}{\omega_o}}{\sqrt{\frac{64\eta_I^2\eta_R^2}{n^2\pi^4M_V^2} - [1 - (\frac{\omega}{\omega_o})^2]^2}}$$

$$= \frac{0.9}{\sqrt{\frac{64\times0.95^2\times0.92^2}{\frac{1}{3^2}\times\pi^4\times5^2} - [1 - 0.9^2]^2}} = 2.37.$$

Using (7.4) and (7.5), the resonant inductance can be found as

$$L = \frac{R_i^2C}{Q_L^2} = \frac{500^2 \times 4.7 \times 10^{-9}}{2.37^2} = 209.7 \ \mu H.$$

16.2 A transformerless half-bridge PRC with a half-wave rectifier supplies 100 W power to a resistance of 25 Ω. The input voltage is $V_I = 200$ V, the normalized switching frequency is $\omega/\omega_o = 0.9$, and the loaded quality factor is $Q_L = 3$. What is the efficiency of the converter?

From (3.30), (16.2) and (16.1), the voltage transfer function of the converter is

$$M_V \equiv \frac{V_O}{V_I} = \frac{2\eta}{\pi^2\sqrt{[1 - (\frac{\omega}{\omega_o})^2]^2 + [\frac{1}{Q_L}(\frac{\omega}{\omega_o})]^2}}.$$

The output voltage of the converter can be found as

$$V_O = \sqrt{P_OR_L} = \sqrt{100 \times 25} = 50 \text{ V}.$$

Hence, the efficiency of the converter is

$$\eta = \frac{V_O \pi^2 \sqrt{[1 - (\frac{\omega}{\omega_o})^2]^2 + [\frac{1}{Q_L}(\frac{\omega}{\omega_o})]^2}}{2V_I}$$

$$= \frac{50\pi^2 \sqrt{[1 - 0.9^2]^2 + (\frac{0.9}{3})^2}}{2 \times 200} = 43.81\%.$$

16.3 Design a full-bridge parallel resonant converter with a transformer center-tapped rectifier to meet the following specifications: $V_I = 400$ V, $V_O = 180$ V, $R_{Lmin} = 125$ Ω, and the switching frequency $f = 180$ kHz. Assume the total efficiency of the converter at full load $\eta = 90\%$, the rectifier efficiency $\eta_R = 97\%$, the transformer turns ratio $n = 4$, and the corner frequency of the inverter $f_o = 200$ kHz.

The maximum output power of the converter is

$$P_{Omax} = \frac{V_O^2}{R_{Lmin}} = \frac{180^2}{125} = 259.2 \text{ W}$$

and the maximum value of the dc load current is

$$I_{Omax} = \frac{V_O}{R_{Lmin}} = \frac{180}{125} = 1.44 \text{ A}.$$

The maximum dc input power can be calculated as

$$P_{Imax} = \frac{P_{Omax}}{\eta} = \frac{259.2}{0.9} = 288 \text{ W}$$

and the maximum value of the dc input current

$$I_{Imax} = \frac{P_{Imax}}{V_I} = \frac{288}{400} = 0.72 \text{ A}.$$

Using (3.66), one obtains the equivalent input resistance of the rectifier

$$R_i = \frac{\pi^2 n^2 R_L}{8\eta_R} = \frac{\pi^2 \times 4^2 \times 125}{8 \times 0.97} = 2544 \ \Omega.$$

The voltage transfer function of the rectifier can be found from (3.66) and (3.67) as

$$M_{VR} = \frac{2\sqrt{2}\eta_R}{n\pi} = \frac{2\sqrt{2} \times 0.97}{4 \times \pi} = 0.2183.$$

According to (3.58) and (3.13), the diode peak voltage V_{DM} and current I_{DM} are

$$V_{DM} = \pi V_O = \pi \times 180 = 565.5 \text{ V}$$

and

$$I_{DM} = I_O = 1.44 \text{ A}.$$

The dc-to-dc voltage transfer function of the converter is $M = V_O/V_I = 180/400 = 0.45$. The required efficiency of the inverter is $\eta_I = \eta/\eta_R = 0.9/0.97 = 0.9278$. Equations (7.35), (7.97), (7.46), and (3.67) produce

$$M = \eta_I M_{Vs} M_{Vr} M_{VR}$$

from which

$$M_{Vr} = \frac{M}{\eta_I M_{Vs} M_{VR}} = \frac{0.45}{0.9278 \times 0.9 \times 0.2183} = 2.469.$$

Rearranging (7.35), the loaded quality factor Q_L can be calculated as

$$Q_L = \frac{\omega/\omega_o}{\sqrt{\frac{1}{M_{Vr}^2} - [1 - (\frac{\omega}{\omega_o})^2]^2}} = \frac{0.9}{\sqrt{\frac{1}{2.469^2} - (1 - 0.9^2)^2}} = 2.52.$$

The values of the resonant components are

$$L = \frac{R_i}{\omega_o Q_L} = \frac{2544}{2 \times \pi \times 200 \times 10^3 \times 2.52} = 803.4 \ \mu\text{H}$$

and

$$C = \frac{Q_L}{\omega_o R_i} = \frac{2.52}{2 \times \pi \times 200 \times 10^3 \times 2544} = 788 \text{ pF}.$$

The characteristic impedance of the resonant circuit is

$$Z_o = \sqrt{\frac{L}{C}} = \sqrt{\frac{803.4 \times 10^{-6}}{788 \times 10^{-12}}} = 1010 \ \Omega.$$

Hence, (7.48) gives

$$I_m = I_{SM} = \frac{4 V_I M_{Vr} \sqrt{1 + (Q_L \frac{\omega}{\omega_o})^2}}{\pi Z_o Q_L}$$

$$= \frac{4 \times 400 \times 2.469 \sqrt{1 + (2.52 \times 0.9)^2}}{\pi \times 1010 \times 2.52} = 1.225 \text{ A}.$$

Chapter 17

17.1 A transformerless half-bridge SPRC with a half-wave rectifier has the following parameters: the output power $P_O = 50$ W, the load resistance $R_L = 50$ Ω, the input voltage $V_I = 280$ V, the normalized switching frequency $\omega/\omega_o = 0.9$, the loaded quality factor $Q_L = 0.3$, and the ratio of capacitances $A = 1.2$. What is the efficiency of the converter?

The output voltage of the converter is

$$V_O = \sqrt{P_O R_L} = \sqrt{50 \times 50} = 50 \text{ V}.$$

From (17.2), (17.1), and (3.30), the efficiency of the converter can be obtained as

$$\eta = \frac{V_O \pi^2 \sqrt{(1+A)^2[1-(\frac{\omega}{\omega_o})^2]^2 + [\frac{1}{Q_L}(\frac{\omega}{\omega_o} - \frac{\omega_o}{\omega}\frac{A}{A+1})]^2}}{2V_I}$$

$$= \frac{50 \times \pi^2 \sqrt{(1+1.2)^2(1-0.9^2)^2 + [\frac{1}{0.3}(0.9 - \frac{1}{0.9}\frac{1.2}{1.2+1})]^2}}{2 \times 280} = 93.87\%.$$

17.2 A full-bridge SPRC with a transformer center-tapped rectifier supplies power to a 40-Ω load resistance. The converter efficiency is $\eta = 91\%$, the efficiency of the inverter is $\eta_I = 96\%$, and the transformer turns ratio is $n = 2$. What is the resonant inductance of the inverter if the equivalent capacitance is $C = 1$ nF and the loaded quality factor is $Q_L = 0.4$?

Using (17.3), the efficiency of the rectifier is

$$\eta_R = \frac{\eta}{\eta_I} = \frac{0.91}{0.96} = 94.79\%.$$

From (3.66), the input resistance of the rectifier can be calculated as

$$R_i = \frac{\pi^2 n^2 R_L}{8\eta_R} = \frac{\pi^2 \times 2^2 \times 40}{8 \times 0.9479} = 208.2 \text{ Ω}.$$

Equations (8.4) and (8.5) give the value of the resonant inductance as

$$L = \frac{R_i^2 C}{Q_L^2} = \frac{208.2^2 \times 10^{-9}}{0.4^2} = 270.9 \ \mu\text{H}.$$

17.3 Design a full-bridge SPRC with a transformer center-tapped rectifier. The following specifications should be met: input voltage V_I=270 V, output voltage $V_O = 48$ V, output current $I_O = 0$ to 4 A, and operating frequency $f = 200$ kHz. Assume that the efficiency of the converter is $\eta = 90\%$, the efficiency of the inverter is $\eta_I = 93\%$, the ratio of capacitances is $A = 1$, the normalized switching frequency is $f/f_o = 0.95$, and the transformer turns ratio is $n = 4$.

From (17.3), the efficiency of the rectifier is

$$\eta_R = \frac{\eta}{\eta_I} = \frac{0.9}{0.93} = 96.77\%.$$

The maximum output power is

$$P_{Omax} = V_O I_{Omax} = 48 \times 4 = 192 \text{ W}.$$

The full–load resistance of the converter can be calculated as

$$R_{Lmin} = \frac{V_O}{I_{Omax}} = \frac{48}{4} = 12 \ \Omega.$$

The maximum dc input power is

$$P_{Imax} = \frac{P_{Omax}}{\eta} = \frac{192}{0.9} = 213.3 \text{ W}$$

and the maximum value of the dc input current is

$$I_{Imax} = \frac{P_{Imax}}{V_I} = \frac{213.3}{270} = 0.79 \text{ A}.$$

Equation (3.66) gives the equivalent input resistance of the rectifier at full load as

$$R_{imin} = \frac{\pi^2 n^2 R_L}{8\eta_R} = \frac{\pi^2 \times 4^2 \times 12}{8 \times 0.9677} = 244.8 \ \Omega.$$

The voltage transfer function of the rectifier can be found from (3.67) as

$$M_{VR} = \sqrt{\frac{\eta_R R_L}{R_i}} = \sqrt{\frac{0.9677 \times 12}{244.8}} = 0.2178.$$

The voltage transfer function of the converter is

$$M_V = \frac{V_O}{V_I} = \frac{48}{270} = 0.1778.$$

Thus, using (8.66), the required transfer function of the resonant circuit is

$$M_{Vr} = \frac{M_V}{\eta M_{Vs} M_{VR}} = \frac{0.1778}{0.9 \times 0.9 \times 0.2178} = 1.01.$$

Using (8.25), the loaded quality factor can be obtained as

$$Q_L = \frac{\frac{\omega}{\omega_o} - \frac{\omega_o}{\omega} \frac{A}{A+1}}{\sqrt{\frac{1}{M_{Vr}^2} - (1 + A)^2 [1 - (\frac{\omega}{\omega_o})^2]^2}}$$

$$= \frac{0.95 - \frac{1}{0.95 \times 2}}{\sqrt{\frac{1}{1.01^2} - 2^2 (1 - 0.95^2)^2}} = 0.436.$$

The resonant frequency is $f_o = f/(f/f_o) = 200/0.95 = 210.5$ kHz. From (8.5) and (8.2), the values of the resonant components are

$$L = \frac{R_{imin}}{\omega_o Q_L} = \frac{244.8}{2 \times \pi \times 210.5 \times 10^3 \times 0.436} = 424.5 \ \mu\text{H}$$

$$C = \frac{Q_L}{\omega_o R_{imin}} = \frac{0.436}{2 \times \pi \times 210.5 \times 10^3 \times 244.8} = 1.35 \text{ nF}$$

$$C_1 = C(1 + \frac{1}{A}) = 2C = 2.7 \text{ nF}$$

$$C_2 = C(1 + A) = 2C = 2.7 \text{ nF}.$$

From (8.69), the maximum switch current equal to the amplitude of the current through the resonant circuit is

$$I_{SM} = I_m = \frac{4V_I}{\pi R_{imin}} \sqrt{\frac{1 + [Q_L(\frac{\omega}{\omega_o})(1 + A)]^2}{(1 + A)^2 [1 - (\frac{\omega}{\omega_o})^2]^2 + \frac{1}{Q_L^2}(\frac{\omega}{\omega_o} - \frac{\omega_o}{\omega} \frac{A}{A+1})^2}}$$

$$= \frac{4 \times 270}{\pi \times 244.8} \sqrt{\frac{1 + [0.436 \times 0.95(1 + 1)]^2}{(1 + 1)^2 (1 - 0.95^2)^2 + \frac{1}{0.436^2}(0.95 - \frac{1}{0.95} \times \frac{1}{1+1})^2}}$$

$$= 1.84 \text{ A}.$$

The peak values of the voltages across the reactive components can be obtained from (8.70), (8.71), and (8.72) as

$$V_{Lm} = \omega L I_m = 2\pi \times 200 \times 10^3 \times 424.5 \times 10^{-6} \times 1.84 = 981.5 \text{ V}$$

$$V_{C1m} = \frac{I_m}{\omega C_1} = \frac{1.84}{2\pi \times 200 \times 10^3 \times 2.7 \times 10^{-9}} = 542.3 \text{ V}$$

and

$$V_{C2m} = \frac{4V_I}{\pi\sqrt{(1+A)^2[1-(\frac{\omega}{\omega_o})^2]^2 + [\frac{1}{Q_L}(\frac{\omega}{\omega_o} - \frac{\omega_o}{\omega}\frac{A}{A+1})]^2}}$$

$$= \frac{4 \times 270}{\pi\sqrt{(1+1)^2(1-0.95^2)^2 + [\frac{1}{0.436}(0.95 - \frac{1}{0.95} \times \frac{1}{1+1})]^2}} = 346.9 \text{ V}.$$

Chapter 18

18.1 A transformerless full-bridge CLL RC with a half-wave rectifier delivers 121 W power to 25 Ω load resistance with an efficiency of 89%. The inverter operates at the normalized operating frequency $f/f_o = 1.15$. What is the characteristic impedance of the inverter if the input voltage of the converter is $V_I = 280$ V, the inductance ratio is $A = 1$, and the efficiency of the rectifier is $\eta_R = 96\%$?

Using (3.31), the input resistance of the rectifier can be obtained as

$$R_i = \frac{\pi^2 R_L}{2\eta_R} = \frac{\pi^2 \times 25}{2 \times 0.96} = 128.5 \ \Omega.$$

The output voltage of the converter can be found as

$$V_O = \sqrt{P_O R_L} = \sqrt{121 \times 25} = 55 \ \text{V}.$$

From (18.24) and (18.1), the voltage transfer function of the converter is

$$M_V \equiv \frac{V_O}{V_I} = \frac{4\eta}{\pi^2 \sqrt{(1+A)^2[1-(\frac{\omega_o}{\omega})^2]^2 + [\frac{1}{Q_L}(\frac{\omega}{\omega_o}\frac{A}{A+1} - \frac{\omega_o}{\omega})]^2}}.$$

Hence, the loaded quality factor is

$$Q_L = \frac{\left|\frac{\omega}{\omega_o}\frac{A}{A+1} - \frac{\omega_o}{\omega}\right|}{\sqrt{\frac{16\eta^2 V_I^2}{\pi^4 V_O^2} - (1+A)^2[1-(\frac{\omega_o}{\omega})^2]^2}}$$

$$= \frac{\left|1.15\frac{1}{1+1} - \frac{1}{1.15}\right|}{\sqrt{\frac{16 \times 0.89^2 \times 280^2}{\pi^4 \times 55^2} - (1+1)^2[1-(\frac{1}{1.15})^2]^2}} = 0.1664.$$

Using (9.5), the characteristic impedance of the resonant circuit can be obtained as

$$Z_o = \frac{R_i}{Q_L} = \frac{128.5}{0.1664} = 772.2 \ \Omega.$$

18.2 A half-bridge CLL RC with a bridge rectifier has the following parameters: input voltage $V_I = 100$ V, output voltage $V_O = 360$ V, and operating frequency $f = 150$ kHz. The loaded quality factor of the inverter is $Q_L = 0.5$, the resonant capacitance is $C = 4.7$ nF, and the resonant inductances are $L_1 = 250$ μH and $L_2 = 200$ μH. The transformer turns ratio is $n = 1/5$ and the efficiency of the rectifier is $\eta_R = 96\%$. Calculate the input power of the converter.

From (9.1) and (9.2), the ratio of inductances is

$$A = \frac{L_1}{L_2} = \frac{250}{200} = 1.25$$

and the equivalent inductance is

$$L = L_1 + L_2 = 250 + 200 = 450 \ \mu\text{H}.$$

Equation (9.3) gives the corner frequency of the inverter

$$f_o = \frac{1}{2\pi\sqrt{LC}} = \frac{1}{2\pi\sqrt{450 \times 10^{-6} \times 4.7 \times 10^{-9}}} = 109.4 \text{ kHz}.$$

Using (9.5) and (9.4), the input resistance of the rectifier can be calculated as

$$R_i = Q_L\sqrt{\frac{L}{C}} = 0.5\sqrt{\frac{450 \times 10^{-6}}{4.7 \times 10^{-9}}} = 154.7 \ \Omega.$$

From (3.81), the dc load resistance of the converter is

$$R_L = \frac{8\eta_R R_i}{\pi^2 n^2} = \frac{8 \times 0.96 \times 154.7}{\frac{\pi^2}{5^2}} = 3009.6 \ \Omega.$$

Hence, the output power is

$$P_O = \frac{V_O^2}{R_L} = \frac{360^2}{3009.6} = 43.1 \text{ W}.$$

Using (18.1) and (18.24), one obtains the efficiency of the converter as

$$\eta = \frac{n\pi^2 V_0\sqrt{(1+A)^2[1-(\frac{f_o}{f})^2]^2 + [\frac{1}{Q_L}(\frac{f}{f_o}\frac{A}{A+1} - \frac{f_o}{f})]^2}}{4V_I}$$

$$= \frac{\frac{1}{5}\pi^2 \times 180\sqrt{(1+1.25)^2[1-(\frac{109.4}{150})^2]^2 + [\frac{1}{0.5}(\frac{150}{109.4}\frac{1.25}{1.25+1} - \frac{109.4}{150})]^2}}{4 \times 48}$$

$$= 93.73\%.$$

Thus, the input power is

$$P_I = \frac{P_O}{\eta} = \frac{43.1}{0.9373} = 46 \text{ W}.$$

18.3 Design a transformerless full-bridge CLL RC with a half-wave rectifier that meets the following specifications: $V_I = 200$ V, $V_O = 50$ V, and $I_O = 0$ to 4 A. Neglect losses due to the ripple current in the rectifier. Assume the ratio of inductances $A = 0.5$, the resonant frequency $f = 100$ kHz, the normalized operating frequency $f/f_o = 1.41$, the efficiency of the rectifier $\eta_R = 96\%$, and the efficiency of the inverter $\eta_I = 94\%$.

It is sufficient to design the converter for the full power. The maximum output power is

$$P_O = V_O I_{Omax} = 50 \times 4 = 200 \text{ W}.$$

The full-load resistance of the converter is

$$R_L = \frac{V_O}{I_{Omax}} = \frac{50}{4} = 12.5 \ \Omega.$$

The total efficiency of the converter is

$$\eta = \eta_I \eta_R = 0.94 \times 0.96 = 90.24\%.$$

Hence, one obtains the maximum dc input power

$$P_I = \frac{P_O}{\eta} = \frac{200}{0.9024} = 221.6 \text{ W}$$

and the maximum value of the dc input current

$$I_I = \frac{P_I}{V_I} = \frac{221.6}{200} = 1.11 \text{ A}.$$

Equation (3.31) gives the equivalent input resistance of the rectifier at full load

$$R_i = \frac{\pi^2 R_L}{2\eta_R} = \frac{\pi^2 \times 12.5}{2 \times 0.96} = 64.3 \ \Omega.$$

The voltage transfer function of the converter is

$$M_V = \frac{V_O}{V_I} = \frac{50}{200} = 0.25.$$

Rearrangement of (18.24) leads to the loaded quality factor

$$Q_L = \frac{\left| \frac{f}{f_o} \frac{A}{A+1} - \frac{f_o}{f} \right|}{\sqrt{\frac{16\eta^2}{\pi^4 M_V^2} - (1+A)^2 [1 - (\frac{f_o}{f})^2]^2}}$$

$$= \frac{\left| 1.41 \frac{0.5}{0.5+1} - \frac{1}{1.41} \right|}{\sqrt{\frac{16 \times 0.9024^2}{\pi^4 \times 0.25^2} - (1+0.5)^2 [1 - (\frac{1}{1.41})^2]^2}} = 0.19.$$

The values of the resonant components can be calculated using (9.2)–(9.5) as

$$L = \frac{R_i}{\omega_o Q_L} = \frac{64.3}{2 \times \pi \times 100 \times 10^3 \times 0.19} = 538.6 \ \mu\text{H}$$

$$C = \frac{Q_L}{\omega_o R_i} = \frac{0.19}{2 \times \pi \times 100 \times 10^3 \times 64.3} = 4.7 \ \text{nF}$$

$$L_1 = \frac{L}{1 + \frac{1}{A}} = 179.5 \ \mu\text{H}$$

$$L_2 = \frac{L}{1 + A} = 359.1 \ \mu\text{H}.$$

The characteristic impedance of the resonant circuit is $Z_o = R_i/Q_L = 338.4 \ \Omega$.

Chapter 19

19.1 A current-source resonant converter with a bridge rectifier supplies power of 150 W to a load resistance. The parameters of the converter are: input voltage $V_I = 48$ V, output voltage $V_O = 200$ V, loaded quality factor $Q_L = 3.5$, normalized resonant frequency $\omega/\omega_o = 0.93$, efficiency of the resonant circuit $\eta_{rc} = 98\%$, and transformer turns ratio $n = 1/2$. What is the input power of the converter?

From (19.5) and (19.6), the voltage transfer function of the converter can be expressed as

$$M_V \equiv \frac{V_O}{V_I} = \frac{2n\sqrt{1 + [Q_L(\frac{\omega}{\omega_o} - \frac{\omega_o}{\omega})]^2}}{n\eta_{rc}}.$$

Hence, the efficiency of the converter is

$$\eta = \frac{V_O n \eta_{rc}}{2V_I\sqrt{1 + [Q_L(\frac{\omega}{\omega_o} - \frac{\omega_o}{\omega})]^2}}$$

$$= \frac{200 \times \frac{1}{2} \times 0.98}{2 \times 48\sqrt{1 + [3.5(0.93 - \frac{1}{0.93})]^2}} = 91\%$$

and the input power is

$$P_I = \frac{P_O}{\eta} = \frac{150}{0.91} = 164.8 \text{ W}.$$

19.2 A current-source resonant converter with a half-wave rectifier has the following parameters: input voltage $V_I = 180$ V, output voltage $V_O = 100$ V, load resistance $R_L = 100$ Ω, resonant frequency $f_o = 100$ kHz, efficiency of the inverter $\eta_I = 95\%$, efficiency of the resonant circuit $\eta_{rc} = 99\%$, efficiency of the rectifier $\eta_R = 95\%$, resonant inductance $L = 1$ mH, transformer turns ratio $n = 2$, and operation below the resonant frequency. Calculate the operating frequency of the converter.

From (3.31), the input resistance of the rectifier is

$$R_i = \frac{\pi^2 n^2 R_L}{2\eta_R} = \frac{\pi^2 \times 2^2 \times 100}{2 \times 0.95} = 2078 \text{ } \Omega.$$

Equations (11.9) and (11.14) give the loaded quality factor as

$$Q_L = \frac{R}{\omega_o L} = \frac{\eta_{rc} R_i}{\omega_o L} = \frac{0.99 \times 2078}{2\pi \times 100 \times 10^3 \times 10^{-3}} = 3.27.$$

Using (19.1) and (19.2), the voltage transfer function of the converter is obtained as

$$M_V \equiv \frac{V_O}{V_I} = \frac{\eta_I \eta_R \sqrt{1 + [Q_L(\frac{f}{f_o} - \frac{f_o}{f})]^2}}{n \eta_{rc}}.$$

Thus,

$$\left| \frac{f}{f_o} - \frac{f_o}{f} \right| = \frac{1}{Q_L} \sqrt{\left(\frac{V_O n \eta_{rc}}{V_I \eta_I \eta_R} \right)^2 - 1}$$

$$= \frac{1}{3.27} \sqrt{\left(\frac{100 \times 2 \times 0.99}{180 \times 0.95 \times 0.95} \right)^2 - 1} = 0.2131$$

from which $f/f_o = 0.899$ and

$$f = 89.9 \text{ kHz}.$$

19.3 Design a current-source converter with a bridge rectifier to meet the following specifications: $V_I = 48$ V, $V_O = 280$ V, and $P_{Omax} = 100$ W. Assume the total converter efficiency $\eta = 90\%$, the efficiency of the rectifier $\eta_R = 94\%$, the efficiency of the resonant circuit $\eta_{rc} = 99\%$, the transformer turns ratio $n = 1/3$, the normalized switching frequency $\omega/\omega_o = 0.95$, and the resonant frequency $f_o = 100$ kHz. Neglect losses due to the ripple current in the rectifier.

It is sufficient to design the converter for the full power. The maximum dc input power is

$$P_I = \frac{P_{Omax}}{\eta} = \frac{100}{0.9} = 111.1 \text{ W}$$

and the maximum value of the dc input current is

$$I_I = I_{SM} = \frac{P_I}{V_I} = \frac{111.1}{48} = 2.315 \text{ A}.$$

The full-load resistance of the converter can be calculated as

$$R_L = \frac{V_O^2}{P_{Omax}} = \frac{280^2}{100} = 784 \ \Omega.$$

The maximum value of the dc load current is

$$I_O = \frac{V_O}{R_L} = \frac{280}{784} = 0.357 \text{ A.}$$

Using (3.81), one obtains the equivalent input resistance of the rectifier

$$R_i = \frac{\pi^2 n^2 R_L}{8 \eta_R} = \frac{\pi^2 \times \frac{1}{3^2} \times 784}{8 \times 0.94} = 114.3 \text{ }\Omega.$$

The voltage transfer function of the rectifier can be found from (3.82)

$$M_{VR} = \sqrt{\frac{\eta_R R_L}{R_i}} = \sqrt{\frac{0.94 \times 784}{114.3}} = 2.54.$$

From (3.13) and (3.83), the diode peak current I_{DM} and voltage V_{DM} are

$$I_{DM} = I_O = 0.357 \text{ A}$$

$$V_{DM} = \frac{\pi}{2} V_O = \frac{\pi}{2} \times 280 = 439.8 \text{ V.}$$

The dc–to–dc voltage transfer function of the converter is

$$M_V = \frac{V_O}{V_I} = \frac{280}{48} = 5.833.$$

The required efficiency of the inverter is

$$\eta_I = \frac{\eta}{\eta_R} = \frac{0.90}{0.94} = 0.9574.$$

Rearranging (19.5), the loaded quality factor is

$$Q_L = \frac{\sqrt{\left(\frac{n M_V \eta_{rc}}{2\eta}\right)^2 - 1}}{\left| \frac{\omega}{\omega_o} - \frac{\omega_o}{\omega} \right|} = \frac{\sqrt{\left(\frac{\frac{1}{3} \times 5.833 \times 0.99}{2 \times 0.9}\right)^2 - 1}}{\left| 0.95 - \frac{1}{0.95} \right|} = 3.69.$$

The total resistance of the inverter is $R = \eta_{rc} R_i = 0.99 \times 114.3 = 113.2 \text{ }\Omega.$ The values of the resonant component are

$$L = \frac{R}{\omega_o Q_L} = \frac{113.2}{2 \times \pi \times 100 \times 10^3 \times 3.69} = 48.8 \text{ }\mu\text{H}$$

$$C = \frac{Q_L}{\omega_o R} = \frac{3.69}{2 \times \pi \times 100 \times 10^3 \times 113.2} = 51.9 \text{ nF.}$$

The characteristics impedance is $Z_o = \sqrt{L/C} = 30.7\ \Omega$. Equation (11.17) gives the maximum amplitude of the input current to the resonant circuit $I_m = (2/\pi)I_{Imax} = 1.47$ A. The maximum switch voltage V_{SM} is equal to the peak value of the voltage across the resonant circuit. Therefore, from (11.24),

$$V_{SM} = \sqrt{2}V_{Ri} = \pi V_I \frac{\eta_I}{\eta_{rc}}\sqrt{1 + [Q_L(\frac{\omega}{\omega_o} - \frac{\omega_o}{\omega})]^2}$$

$$= \pi \times 48 \times \frac{0.9574}{0.99}\sqrt{1 + [3.69(0.95 - \frac{1}{0.95})]^2} = 155.9 \text{ V}.$$

Chapter 20

20.1 A rectifier in a Class D-E resonant converter of Fig. 20.1(a) operates with a duty ratio $D = 0.5$. The parameters of the inverter are: input voltage $V_I = 200$ V, efficiency $\eta_I = 96\%$, loaded quality factor $Q_L = 2.85$, and normalized switching frequency $f/f_o = 1.1$. What is the output voltage of the converter?

Rearrangement of (20.8) gives the output voltage as

$$V_O = \frac{V_I \eta_I \sqrt{\pi^2 + 4}}{2\pi \sqrt{1 + Q_L^2(\frac{\omega}{\omega_o} - \frac{\omega_o}{\omega})^2}} = \frac{200 \times 0.96\sqrt{\pi^2 + 4}}{2\pi\sqrt{1 + 2.85^2(1.1 - \frac{1}{1.1})^2}} = 100 \text{ V}.$$

20.2 A Class D-E resonant converter operating with a switching frequency 200 kHz supplies 100 W power at a 50 V output voltage. The inverter has the following parameters: input voltage $V_I = 100$ V, efficiency $\eta_I = 94\%$, loaded quality factor $Q_L = 3$, and resonant frequency $f_o = 180$ kHz. Calculated the approximate value of the shunt capacitor C_2 in the rectifier.

The dc load resistance of the converter is

$$R_L = \frac{V_O^2}{P_O} = \frac{50^2}{100} = 25 \ \Omega.$$

The approximate value of the shunt capacitor C_2 can be obtained from (20.14) as

$$C_2 = \frac{\frac{V_O \pi \sqrt{1 + Q_L^2(\frac{f}{f_o} - \frac{f_o}{f})^2}}{\sqrt{2}V_I \eta_I} - 1}{0.8\omega R_L}$$

$$= \frac{\frac{50\pi\sqrt{1 + 3^2(\frac{200}{180} - \frac{180}{200})^2}}{\sqrt{2}\times 100 \times 0.94} - 1}{0.8 \times 2 \times \pi \times 200 \times 25} = 15.9 \text{ nF}.$$

20.3 Design the Class D-E converter of Fig. 20.1(a) to meet the following specifications: input voltage $V_I = 200$ V, output voltage $V_O = 100$ V, and load resistance $R_L = 50 \ \Omega$ to ∞. Assume the resonant frequency $f_o = 100$ kHz, the normalized switching frequency $f/f_o = 1.07$, the converter efficiency $\eta = 90\%$, and the inverter efficiency $\eta_I = 95\%$.

The maximum value of the dc output current is

$$I_{Omax} = \frac{V_O}{R_{Lmin}} = \frac{100}{50} = 2 \text{ A}$$

and the maximum value of the dc output power is

$$P_{Omax} = V_O I_{Omax} = 100 \times 2 = 200 \text{ W}.$$

The rectifier will be designed for $D = 0.5$ which gives maximum power-output capability (the best utilization of the rectifier's diode). From (20.2) to (20.7), the parameters of the Class E rectifier are:

$$C_2 = \frac{1}{2\pi^2 f R_{Lmin}} = \frac{1}{2\pi^2 \times 107 \times 10^3 \times 50} = 9.47 \text{ nF}$$

$$R_i = \frac{8R_L}{\pi^2 + 4} = 0.5769 \times 50 = 28.84 \text{ } \Omega$$

$$C_i = \frac{2(\pi^2 + 4)C_2}{\pi^2 - 4} = 4.7259 \times 9.65 \times 10^{-9} = 44.75 \text{ nF}$$

$$I_{DM} = (\frac{\sqrt{\pi^2 + 4}}{2} + 1)I_{Omax} = 2.862 \times 2 = 5.724 \text{ A}$$

and

$$V_{DM} = 2\pi arctan \left(\frac{2}{\pi}\right) V_O = 3.562 \times 100 = 356.2 \text{ V}.$$

The voltage transfer function of the converter is

$$M_V = \frac{V_O}{V_I} = \frac{100}{200} = 0.5.$$

Thus, using (20.8) and (20.1), the loaded quality factor is

$$Q_L = \frac{\sqrt{\frac{\eta_I^2(\pi^2+4)}{4\pi^2 M_V^2} - 1}}{|\frac{\omega}{\omega_o} - \frac{\omega_o}{\omega}|} = \frac{\sqrt{\frac{0.95^2(\pi^2+4)}{4\pi^2 \times 0.5^2} - 1}}{|1.07 - \frac{1}{1.07}|} = 3.82.$$

The parameters of the inverter are

$$L_1 = \frac{Q_L R_i}{\omega_o} = \frac{3.82 \times 28.84}{2\pi \times 100 \times 10^3} = 175.3 \text{ } \mu H$$

$$C = \frac{1}{\omega_o Q_L R_i} = \frac{1}{2\pi \times 100 \times 10^3 \times 3.82 \times 28.84} = 14.45 \text{ nF}$$

$$C_1 = \frac{CC_i}{C_i - C} = \frac{14.45 \times 44.75}{44.75 - 14.45} = 21.34 \text{ nF}$$

and

$$P_{Imax} = \frac{P_{Omax}}{\eta} = \frac{200}{0.9} = 222.2 \text{ W}.$$

The assumed value of the rectifier efficiency is $\eta_R = \eta/\eta_I = 0.9/0.95 = 0.9474$. Hence, the amplitude of the current through the resonant circuit equal to the maximum switch current is

$$I_m = I_{SM} = \sqrt{\frac{2P_{Omax}}{\eta_R R_i}} = \sqrt{\frac{2 \times 200}{0.9474 \times 28.84}} = 3.83 \text{ A}.$$

The amplitude of the voltage across the resonant capacitor is

$$V_{C1m} = \frac{I_m}{\omega C_1} = \frac{3.83}{2\pi \times 107 \times 10^3 \times 21.34 \times 10^{-9}} = 267 \text{ V}$$

and the amplitude of the voltage across the resonant inductor is

$$V_{L1m} = \omega L_1 I_m = 2\pi \times 107 \times 10^3 \times 175.3 \times 10^{-6} \times 3.83 = 451.4 \text{ V}.$$

Chapter 21

21.1   A transformerless single-capacitor phase-controlled series resonant converter with a half-wave rectifier supplies power of 200 W to a 50-Ω load resistance. The parameters of the inverter are: the input voltage $V_I = 220$ V, the normalized switching frequency $\omega/\omega_o = 1.3$, the loaded quality factor $Q_L = 3$, the efficiency $\eta_I = 94\%$, and $\phi = 30°$. Calculate the efficiency of the converter.

The output voltage of the converter is

$$V_O = \sqrt{P_O R_L} = \sqrt{200 \times 50} = 100 \text{ V}.$$

Using (21.1) and (21.2), the efficiency of the converter can be calculated as

$$\eta = \frac{V_O\sqrt{1 + Q_L^2(\frac{\omega}{\omega_o} - \frac{\omega_o}{\omega})^2}}{V_I cos(\frac{\phi}{2})} = \frac{100\sqrt{1 + 3^2(1.3 - \frac{1}{1.3})^2}}{220 cos(\frac{30}{2})} = 88.5\%.$$

21.2   A single-capacitor phase-controlled series resonant converter with a transformer center-tapped rectifier converts 280 V to 48 V with 91% efficiency. The inverter has the following parameters: switching frequency $f = 200$ kHz, resonant inductance $L = 542.2\ \mu$H, resonant capacitance $C = 4.7$ nF, and $\phi = 25°$. The efficiency of the rectifier is 95% and the transformer turns ratio is $n = 1$. What is the output power of the converter?

The resonant frequency of the converter is given by (12.10)

$$f_o = \frac{\sqrt{2}}{2\pi\sqrt{LC}} = \frac{\sqrt{2}}{2\pi\sqrt{542.2 \times 10^{-6} \times 4.7 \times 10^{-9}}} = 141 \text{ kHz}.$$

From (21.3) and (21.4), the loaded quality factor can be obtained as

$$Q_L = \frac{\sqrt{\frac{n^2 V_I^2 cos^2(\frac{\phi}{2})}{4n^2 V_O^2} - 1}}{\left|\frac{f}{f_o} - \frac{f_o}{f}\right|} = \frac{\sqrt{\frac{0.91^2 \times 280^2 cos^2(\frac{25}{2})}{4 \times 1^2 \times 48^2} - 1}}{\left|\frac{200}{141} - \frac{141}{200}\right|} = 3.35.$$

Substituting (2.72) into (12.11) and solving the resulting equation with respect to R_L, one obtains

$$R_L = \frac{\pi^2 \omega_o L \eta_R}{16n^2 Q_L}$$

$$= \frac{\pi^2 \times 2\pi \times 141 \times 10^3 \times 542.2 \times 10^{-6} \times 0.95}{16 \times 1^2 \times 3.35} = 84.03 \ \Omega.$$

Hence, the output power is

$$P_O = \frac{V_O^2}{R_L} = \frac{48^2}{84.03} = 27.42 \ \text{W}.$$

21.3 Design a single-capacitor phase-controlled series resonant converter with a half-wave rectifier. The following specifications should be met: $V_I = 200$ V, $V_O = 28$ V, and $P_{Omax} = 50$ W. Assume the resonant frequency $f_o = 120$ kHz, the normalized switching frequency $f/f_o = 1.25$, the inverter efficiency $\eta_I = 94\%$, the rectifier efficiency $\eta_R = 95\%$, the transformer turns ratio $n = 4$, and $cos(\phi/2) = 0.9$ at full load.

It is sufficient to design the converter for full power. The dc load resistance is

$$R_L = \frac{V_O^2}{P_{Omax}} = \frac{28^2}{50} = 15.68 \ \text{W}.$$

The maximum output current is

$$I_{Omax} = \frac{V_O}{R_L} = \frac{28}{15.68} = 1.79 \ \text{A}.$$

Using (2.28), the input resistance of the rectifier is found to be

$$R_i = \frac{2n^2 R_L}{\pi^2 \eta_R} = \frac{2 \times 4^2 \times 15.68}{\pi^2 \times 0.95} = 53.5 \ \Omega.$$

From (2.31), the voltage transfer function of the rectifier is

$$M_{VR} = \frac{\pi \eta_R}{\sqrt{2}n} = \frac{\pi \times 0.95}{\sqrt{2} \times 4} = 0.5276.$$

Rearrangement of (21.1) gives the maximum required voltage transfer function of the inverter

$$M_{VI} = \frac{V_O}{\eta_I V_I M_{VR}} = \frac{28}{0.94 \times 200 \times 0.5276} = 0.282.$$

Hence, using (12.14), one obtains

$$Q_L = \frac{\sqrt{\frac{2cos^2(\frac{\phi}{2})}{M_{VI}^2\pi^2} - 1}}{\left|\frac{\omega}{\omega_o} - \frac{\omega_o}{\omega}\right|} = \frac{\sqrt{\frac{2\times0.9^2}{0.282^2\pi^2} - 1}}{\left|\ 1.25 - \frac{1}{1.25}\ \right|} = 2.29.$$

From (12.10) and (12.11), the values of the resonant components are

$$L = \frac{2R_iQ_L}{\omega_o} = \frac{2 \times 53.5 \times 2.29}{2 \times \pi \times 120 \times 10^3} = 325 \ \mu H$$

and

$$C = \frac{2}{\omega_o^2 L} = \frac{2}{(2 \times \pi \times 120 \times 10^3)^2 \times 325 \times 10^{-6}} = 10.8 \ nF.$$